Indicators for preserving soil quality

Isabelle Cousin, Maylis Desrousseaux, Sophie Leenhardt, eds

To cite this publication:

Cousin I., Desrousseaux M., Leenhardt S., eds, 2025.
Indicators for preserving soil quality, Versailles, Éditions Quæ, 180 p.

Directorate for Collective Scientific Assessment, Foresight and Advanced Studies (DEPE): Guy Richard, Director.

Scientific leads:

Isabelle Cousin, Research Director, INRAE, Info&Sols Research Unit, Orléans, France.

Maylis Desrousseaux, Associate Professor, Paris School of Urban Planning, Champs-sur-Marne, France.

Project coordination: Sophie Leenhardt, INRAE, DEPE.

This book is a condensed report of the expert extended report authored by the committee of experts, one-off scientific contributors sought by the experts, and DEPE contributors: Isabelle Cousin (eds), Maylis Desrousseaux (eds), Denis Angers, Laurent Augusto, Jean-Sauveur Ay, Adrien Baysse-Lainé, Philippe Branchu, Alain Brauman, Nicolas Chemidlin Prévost-Bouré, Claude Compagnone, Raphaël Gros, Carole Hermon, Catherine Keller, Bertrand Laroche, Germain Meulemans, David Montagne, Guénola Pérès, Nicolas Saby, Emmanuelle Vaudour, Jean Villerd, Cyrille Violle, Virginie Lelièvre, Sybille de Mareschal, Marie-Caroline Brichler, Claire Froger, Julie Itey, Sophie Leenhardt (eds) (2024). Préserver la qualité des sols : vers un référentiel d'indicateurs. Rapport d'étude, INRAE (France). 780 pages – DOI 10.17180/qnpx-x742.

Éditions Quæ
RD 10
78026 Versailles Cedex
www.quae.com / www.quae-open.com

ISBN paper: 978-2-7592-4338-9
ISBN PDF: 978-2-7592-4339-6
ISBN ePub: 978-2-7592-4340-2
ISSN: 2115-1229

Contents

Foreword 5

1. Basis of the study 7

Context and handling of the study commissioning 7

Expert group involved 14

Literature analysed 16

Reading guide and terminology used in this condensed report 18

2. Implementation frameworks for soil quality assessment 23

Actors and mechanisms involved in soil quality assessment 23

Diversity of perceptions of soils and their qualities 25

Co-production of information on soil quality 30

3. Governance frameworks and soil quality 33

Private property and public intervention 33

Measures and economic values of soil quality 38

Criteria used in the field of law 43

Regulating soil de-artificialisation and ecological restoration 50

Territoriality of public intervention on soil quality 54

4. The dimensions of soil quality and health and the choice of functions 59

Historical developments in terminology and concepts 59

A not-yet-stabilised distinction between soil quality and soil health 61

From soil quality to its ecological functions 62

Approaches to assess soil multifunctionality 70

Limitations of a function-based approach 73

5. Assessment process 75

Purpose of the assessment 75

Choice of the *indicandum* and of the relevant indicators 79

Measuring parameter and indicator values 83

Reference framework and interpretation framework 98
Scoring or normalising indicators 101
Multi-criteria aggregation forming a soil quality index 101
Monitoring soil quality over time and space 104

6. Generic list of soil function indicators and assessment trial in a given area 107
Main categories of indicators 107
Selected generic indicators 111
Operationality of indicators 115
Assessing soil quality in France 123

7. Lessons learned and avenues for research 129
Summary of key lessons 129
Areas for research 136

Appendices 141

Bibliographic sources cited 160

Acronyms and abbreviations 172

Working group 176

Foreword

Over the last decade, growing concerns about soil degradation have been accompanied by a proliferation of proposals for indicators and methods of calculation and implementation, from both public and private actors, with varying degrees of scientific rigor. To facilitate consideration of soil quality in public policy evaluation and implementation, the Scientific Interest Group on Soils (GIS Sol) has emphasised the importance of consolidating the available scientific resources available to characterise soil quality, identify and test key indicators that can be used and associated methods.

GIS Sol's mission is to establish and manage the French soil information system in response to requests from public authorities and society. The group of partners it brings together positions it at the interface between public policy and research. To meet the demand for an overview of soil quality indicators, the French Agency for Ecological Transition (Ademe), the French Office for Biodiversity (OFB), the Ministry of Ecological Transition, Energy, Climate and Risk Prevention (MTEECPR), and the Ministry of Agriculture, Food Sovereignty and Forestry (MASAF) commissioned and funded this study. It was conducted by INRAE's Directorate for Collective Scientific Assessment, Foresight and Advanced Studies (DEPE).

The results are published on the INRAE and GIS Sol websites in three document formats available online. The full scientific report includes the background to the assessment, the study's scientific framework, a description of the method used and the cited bibliography, which includes nearly 1,800 references, the analyses produced by the experts, and the general conclusions drawn from them. This condensed report brings together the main findings of the study report, without referencing the entire scientific corpus used. Consequently, the sources listed in the bibliographic appendix to this condensed report do not represent the entire corpus, which can be found in the extended report (Cousin *et al.*, 2024). The summary briefly presents the main lessons learned from this work. In addition, the results were presented at a public symposium on November 20, 2024, video recordings of which are also available online.

NB: The content of this document regarding the Soil Monitoring and Resilience Law is based on the draft version of Directive (EU) 2025/2360, which was adopted after our report was published.

1. Basis of the study

Context and handling of the study commissioning

Context

Soil quality has been a concern since ancient times, long being perceived by predominantly rural populations as essential to their livelihood. Industrialisation and urbanisation have freed human activities from the characteristics and constraints of the soil. This has led to a distancing from the soil that nourishes us, replaced by retail trade and the neutralisation of this surface by more stable and inert coverings. In agriculture, mechanisation, land development and inputs have mitigated the specific characteristics of cultivated land (drainage of wet soils, irrigation of drying soils, tidal flooding and liming of acidic soils, mineral fertilisation of poor soils, etc.). The cultivation of land and increase in yields has coincided with unprecedented demographic growth and changes in dietary habits.

The impact of human activities on soil is now evident in the form of degradation, with the resilience of the soil system limited or even compromised over a period of several decades. Yet soil is the source of many of the resources used by humans, particularly for food production, so its vulnerability to environmental disruption is a cause for concern. Soils play an important role in mitigating these disturbances, but conversely can contribute significantly to them once certain thresholds are exceeded (greenhouse gas emissions associated with the thawing of permafrost–polar soils–, desertification of Mediterranean soils, etc.). The alert has been raised at every institutional level: the overall estimate of the proportion of land affected by degradation is 20 to 40% (United Nations Convention to Combat Desertification, 2022); within the European Union (EU), the European Soil Observatory (EUSO) has found that 60 to 70% of the EU's land area is affected by at least one soil degradation process[1] (artificialisation, erosion, compaction, pollution, excess nutrients, salinisation, loss of organic matter, loss of biodiversity); and in France, around 20,000 ha of natural, agricultural or forest areas are lost each year.[2] Land use changes have therefore been included in the list of planetary boundaries, the exceeding of which compromises the sustainability of conditions for human development.[3]

1. EUSO (European Union Soil Observatory), https://esdac.jrc.ec.europa.eu/esdacviewer/euso-dashboard (accessed on 4/11/2024).

2. French Observatory on Land Artificialisation, https://artificialisation.developpement-durable.gouv.fr (accessed on 9/11/2024).

3. "Planetary boundaries define a safe and just development space for humanity, based on intrinsic biophysical processes that regulate the stability of the Earth and marine systems on a global scale. Nine biophysical processes that together regulate this stability: climate change, biodiversity loss, disruption of nitrogen and phosphorus biogeochemical cycles, land use change, ocean acidification, global water use, stratospheric ozone depletion, increased aerosols in the atmosphere, and the introduction of new entities into the biosphere", Steffen *et al.* (2015; https://doi.org/10.1126/ science.1259855), cited by FRB (2023; https://www.fondationbiodiversite.fr/wp-content/uploads/2023/10/FRB_Prospective_2023.pdff, p. 41).

Soil conservation is closely linked to global issues such as climate and biodiversity. In fact, around 60% of known living species (Anthony *et al.*, 2023) reside in the soil, which plays a major role in the hydrobiogeochemical cycles that regulate the atmospheric content of greenhouse gases. Favourable living conditions for biodiversity and human societies inevitably require soil conservation.

Gradually growing awareness of the role of soils in preserving the ecosystems that support human activities and food security has led to political initiatives and the adoption of institutional instruments at international and national levels. The timeline shown in Figure 1.1 shows the main ones. The mid-2010s were notable in this regard, with important partnerships being established under the auspices of the UN around the International Year of Soils in 2015. Major strategies adopted in the early 2020s at the European level (Green Deal and Pact for Healthy Soil in Europe) and at the French level (National Biodiversity Strategy–SNB 2030, which provides in particular for the development of a national soil strategy) laid the foundations for a soil conservation policy.

On a legislative level, initiatives were taken in the early 2020s in France (Climate and Resilience Law setting a target for limiting land artificialisation) and in the EU (proposal for a framework directive on soil monitoring and resilience).

Soil functioning and human activities

Soils in relation to hydrobiogeochemical cycles and biodiversity

Growing concerns about soil are linked to the fundamental role it plays in the cycles that connect it to the different Earth spheres: the atmosphere, the biosphere, the lithosphere and the hydrosphere. The surface conditions of the soil cover are related to gravity, winds, and the vertical or lateral circulation of water within, below or above the soil cover (surface and deep aquifers, evaporation, evapotranspiration, rainfall, fresh water and sea water). The biosphere influences soils organic matter (decomposing litter, leachate and, above all, roots and soil fauna) as well as by extracting nutrients that the soil makes available to it. In addition to supporting the soil cover, the lithosphere is a source of minerals that are transformed, altered and reorganized through its contributions.

Soil cover is a dynamic flow system. Therefore, it is difficult to discuss soil formation as if soil originated on a specific date (Girard *et al.*, 2023). Some of the processes that govern hydrobiogeochemical cycles constitute inputs to the soil (including weathering of the geological substrate, atmospheric deposits and precipitation, biological activity, and anthropogenic inputs such as fertilisation), while others constitute outputs (including runoff and erosion from the soil surface into watercourses, percolation and leaching into deeper layers, gaseous emissions into the atmosphere, and export through biomass harvesting).

Soils thus play a major role in the cycles of many elements, including carbon, nitrogen, phosphorus and potassium. They are the main global reservoir for some of these elements and the place where they are recycled within organic matter, interacting with microbial, plant and animal organisms. Recent estimates (Friedlingstein *et al.*, 2023) indicate that

Figure 1.1. Timeline of the main institutional initiatives aimed at preserving soil quality, whether directly or indirectly

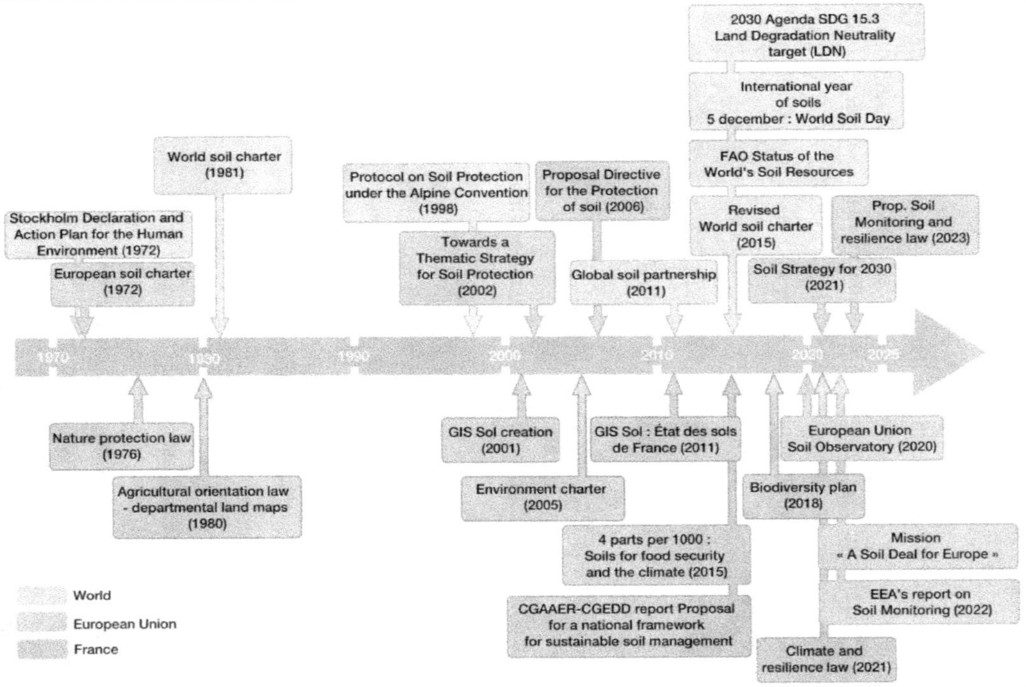

SDGs: Sustainable Development Goals; FAO: Food and Agriculture Organisation of the United Nations; CGAAER: General Council for Food, Agriculture and Rural Areas; CGEDD: General Council for the Environment and Sustainable Development (currently IGEDD); EEA: European Environment Agency; GIS: Scientific Interest Group.

soils are the largest reservoir of organic carbon, surpassing terrestrial biomass, surface sediments and the ocean. Carbon flows between the soil and the atmosphere are significant and can be positive (sequestration) or negative (CO_2 emissions). Increasing, or at least maintaining, soil carbon stocks is considered essential to limiting or mitigating the predicted increase in atmospheric CO_2 and its consequences in terms of climate change (Brunet and Voltz, 2023).

Globally, continental soils receive an average of 800 mm of rainfall per year (Brunet and Voltz, 2023). For France, this average is 600 mm. Soils retain some of the water that infiltrates during wet seasons, which can then be accessed by plants for growth. This water evaporates, either through vegetation or directly from the soil. Human societies have a decisive influence on how rainwater is distributed between infiltration, runoff and percolation into the soil. In temperate environments, the storage capacity of soils can reach 2/5^{e}, or even half of the annual rainfall (Brunet and Voltz, 2023). Unstored water passes through the soil to form groundwater, rivers and lakes, eventually ending up in the seas and oceans.

Soil and human activities: ecosystem services, functions and degradation

Human activities depend on and influence these cycles. The conceptual framework of the ecosystem service cascade proposed by Haines-Young and Potschin (2010) enables us to analyse the relationships between soil properties, its functions and the ecosystem services that benefit humans. Figure 1.2 shows an adapted version of the illustration by Greiner *et al.* (2017). Thus, soil properties strongly determine the benefits that can be expected from it.

Conversely, human interventions in land use planning (e.g. levelling, sealing, drainage) or soil management (e.g. tillage, fertilisation) influence these properties to varying degrees and over different timeframes. This results in soil degradation that can be measured as such. This is the approach that has been favoured to date by the European Soil Observatory,[4] as shown in Figure 1.3. Eight main threats are assessed: artificialisation on the one hand, and, for non-artificialised soils on the other, loss of biodiversity, loss of organic carbon, pollution, excess nutrients, compaction, salinisation and erosion.

Soil is considered degraded when it is artificialised, or when one of these threats reaches a level deemed critical. Some of these processes can occur naturally in soil. For instance, certain contaminants are naturally present in concentrations corresponding to what is known as the geochemical background. It is only when these baseline values, observed in undisturbed environments, are exceeded that the level of contamination caused by human activities is considered. The same applies to erosion and salinisation, which are natural processes of soil evolution but which can be significantly accelerated and aggravated by human activities. It is therefore only above a certain threshold that these threats are considered to be degradation.

4. EUSO (European Union Soil Observatory), https://esdac.jrc.ec.europa.eu/esdacviewer/euso-dashboard (accessed on 4/11/2024)

Figure 1.2. The "cascade" pattern of the relationships between properties, processes, functions, ecosystem services and benefits (based on Greiner *et al.*, 2017)

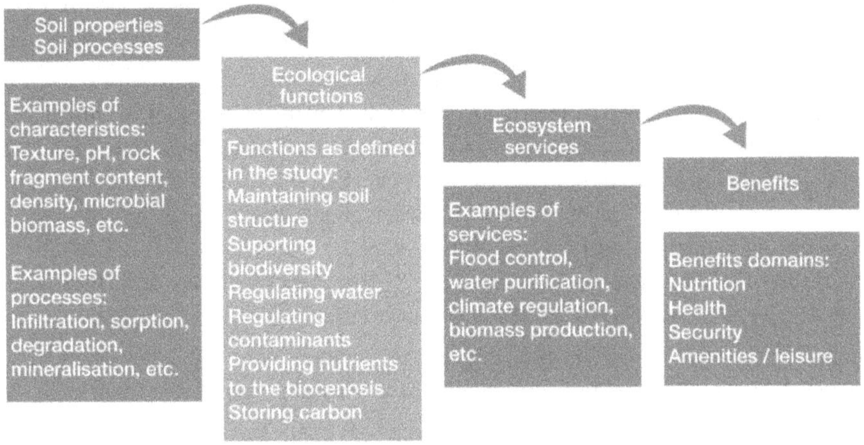

Figure 1.3. Soil degradation in the EU and its coverage.

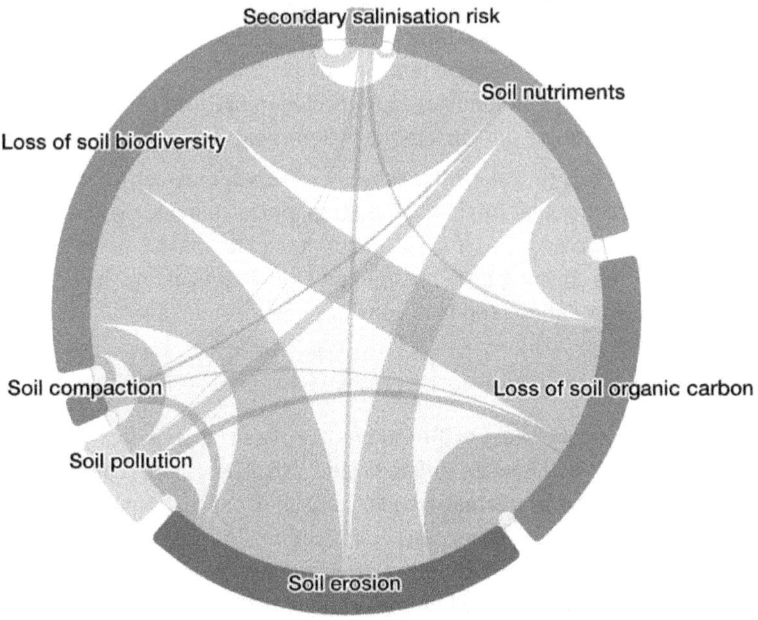

The size of the area connecting two forms of degradation is proportional to the extent of their overlap in hectares (EUSO, European Union Soil Observatory).

Soil, land and land use

Soil is a component of the ecosystem, which also includes vegetation and terrestrial fauna, as well as the ecosociosystem, which incorporates human activities. Depending on the scope, the French terms "*sol*" (soil), "*terre*" (earth) and "*foncier*" (land) refer to overlapping realities.

"Le Sol" (soil) is defined by the French National Soil Science Society (AFES) as "a volume extending from the Earth's surface to a depth marked by the appearance of hard or loose rock, little altered or little affected by pedogenesis. Soil depth can range from a few centimetres to several tens of metres, or more. Locally, it forms part of the soil cover that extends across the entire surface. It most often comprises several horizons corresponding to a specific organisation of organic and/or mineral constituents. This organisation is the result of pedogenesis and the alteration of the parent material. Soil is the site of intense biological activity (roots, fauna and microorganisms)".[5]

Soil can therefore be analysed as a material, as a compartment of the environment, and as a system capable of performing functions and providing ecosystem services.

"La Terre" is the most polysemic of these three terms. It can refer to the planet Earth itself, the material (e.g. in "*terres excavées*" = excavated soil), soil ("*analyses de terre*" = soil test, "*terres noires*" = black soil), or the geographical and land-related dimension included in the English term "land" ("*terres agricoles*" = agricultural land, "*valeur des terres*" = land value). In agricultural terminology, "terre" has a strong land tenure connotation, whereas "*sol*" is more associated with its characteristics (e.g. deep soils, hydromorphic soils). "*Gestion des terres*" (= land management) can evoke the dynamics of land exchanges, while "*gestion des sols*" (= soil management) refers more to a set of agricultural and forestry practices that are considered together in terms of their consequences for the soil.

"Le foncier" (land) refers to real estate, as well as the way in which space is delimited, fitted out, occupied and subject to rights. The geographical dimension is broader, encompassing both the characteristics of the soil, its uses, and its value. The English term **land** can be understood as covering both the land and geographical dimensions in relation to land use, land degradation, and land consumption in "land take".

Identified needs and scope of the study

Alongside the growing awareness of the importance of soil function for human activities, tensions have intensified between different land uses: food production, extraction of materials, renewable energy, recreational spaces, landscapes, housing, infrastructure, industry, waste storage and, in some cases, purification, flood control, etc. Relationships between these uses call for coordination at a broad level and arbitration at a more local level, based on substantiated information that enables soil quality and suitability for intended use to be assessed. In response to this need and in line with advancing

5. https://www.afes.fr/les-sols/definition-et-enjeux (accessed on 31/10/2024).

knowledge, a multitude of scientific and technical proposals have emerged. Commercial offers for soil quality rating services have also arisen, prompting concerns regarding the scientific validity of the sources and methods used, as well as their transparency.

Initial overview of existing approaches

The diversity of interacting variables within the soil and the range of skills required to analyse them give rise to a wide variety of possible approaches. Table A1, in the appendix, lists the main research projects and institutional observatories that have been implemented over the last decade at various levels of governance. These projects differ mainly in terms of the geographical area covered (world, EU, France, sub-national territories), the types of approaches used (soil health, functions or multifunctionality, ecosystem services), their purpose (monitoring, territorial planning, site/plot management), and sometimes even their specific land use (e.g. forestry, agriculture, urban). With the same aim of taking stock of the situation in France, ADEME published in 2023 a review of research projects on soil multifunctionality that it had supported over the previous 20 years (Vincent and Blanchart, 2023).

The presentation pages of these projects very often state the need for a harmonised and calibrated tool for assessing soil quality. This emphasises the crucial importance of developing a common reference framework for all stakeholders and decision-makers. The framework is expected to align with public policy or private management objectives, to be scientifically sound, and to be operationally practical. However, the proposals available to date remain highly diverse. Soil quality is approached in a variety of ways, with concepts such as quality, health, threats of degradation, safety, functions, ecosystem services and fertility being considered. Projects may focus on specific land uses (agricultural, forestry, urban, natural areas) or they may integrate all uses. The parameters measured almost always cover all areas of physics, chemistry and biology. Some studies on ecosystem services and their assessment process economic data. Projects related to soil artificialisation or forest soil management consider physicochemical data and land use data as *proxies* for soil functionality (e.g. FOR-EVAL, Observatoire national de l'artificialisation (French National Observatory on Land Artificialisation), ARTISOLS and DESTISOL). Recognising the importance of biodiversity and the difficulty of assessing it, some projects have focused more specifically on this dimension (e.g. EJP Soil MINOTAUR, ECOFINDERS and Bio-indicators of soil quality).

A common approach to soil quality for all land uses

It is in this context that the present study was conducted, with the aim of clarifying the scientific basis currently available for supporting soil quality assessment. Notably, the study is aligned with the comprehensive perspective promoted by two ongoing public policy initiatives: the Climate and Resilience Law adopted in France in 2021 and introducing restrictions on soil artificialisation, and the European *Soil Monitoring and Resilience* Directive (proposal published in July 2023) providing for the harmonisation of soil degradation monitoring across the EU as a key step towards soil preservation.

These two initiatives address soil quality in a way that is not specific to each use or type of space (natural, agricultural, forest or urban). With this in mind, the assessment carried out here focuses on scientific resources that are suitable for all uses. Work dealing more specifically with the quality required for agricultural or forestry production, for example, has not been targeted. The aim is to produce a shared overview that can be used by all stakeholders to address the effects of changes in land use and for territorial monitoring.

Polluted sites and soils

This study has not examined the legal framework of polluted sites and soils. The principles underpinning the French policy for managing polluted sites and soils are set out in the texts describing the methodology developed in 2007 (Circular of February 8, 2007) and updated in 2017.[6] The current approach to managing polluted sites is based on assessing health and environmental risks according to how the sites are used. Attention is given to the compatibility of the state of the environment with current and future uses. The need to restore this compatibility in order to control health and environmental risks after site rehabilitation is also taken into account. The management plan focuses on treating sources of pollution, particularly concentrated pollution, and controlling residual pollution, while taking remediation techniques and their costs into account.

Pollution indicators were considered and compiled where relevant for an integrated assessment of soil quality and functions. However, the extensive literature specific to polluted sites and soils was not reviewed.

Expert group involved

The working group was set up and the study was conducted in accordance with the Principles for conducting collective scientific assessments and studies at INRAE (Donnars *et al.*, 2021), which are summarised in Box 1.1.

Isabelle Cousin from INRAE and Maylis Desrousseaux from the Paris School of Urban Planning were responsible for the scientific coordination of the study. A committee of scientific experts comprising 19 researchers and research professors from the public sector (including the pilots) was formed.

Participants were identified based on their disciplinary expertise and publications on the study topics. As illustrated in Figure 1.4, the assembled committee is highly multidisciplinary. Two major areas are represented: soil sciences and human and social sciences. In addition to soil ecology, expertise in ecology was sought to consider concepts and reference frameworks developed for other environments that could be relevant to soils. Finally, skills in data processing and information systems management were also mobilised.

---

6. https://ssp-infoterre.brgm.fr/fr/methodologie/methodologie-nationale-gestion-ssp (accessed on 31/10/2024).

Box 1.1. Study principles.

This study is based on a critical analysis of the available worldwide scientific knowledge on the various aspects of soil quality. This literature was analysed by a panel of scientific experts from public research or higher education institutions. As well as summarising perceptions and definitions of soil quality, the study identifies the main stages in assessing soil quality, selecting the key indicators and methods for evaluating soil functions. It also highlights areas of concern, controversial issues, and the dynamics of innovation in this field. By updating the scope of acquired knowledge, and identifying areas of uncertainty, controversy, and knowledge gaps, this work aims to inform various stakeholders' reflections on considering soil quality in public policy. It thus contributes to the mission of research institutes to support public policy.

Experts are selected based on their publications in peer-reviewed scientific journals. Attention is given to the diversity of links of interest (e.g., funding, intellectual affinities and collaborative links), which are inevitable in applied research. Cases of conflict of interest are excluded. Transparency is ensured by describing the sources used and the method used in the study report.

The study is conducted in consultation with the funders (the monitoring committee) and the main parties involved in the subject (the advisory committee of stakeholders).

Figure 1.4. Disciplinary fields of the 19 experts in the study according to the INRAE reference framework (up to 3 macro-disciplines per expert).

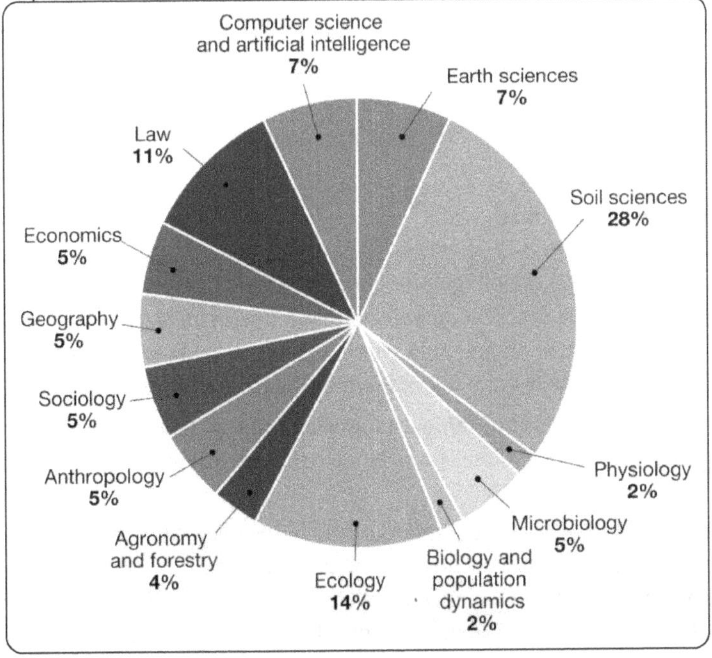

Figure 1.5. Affiliation of the 19 study experts by type of organisation.

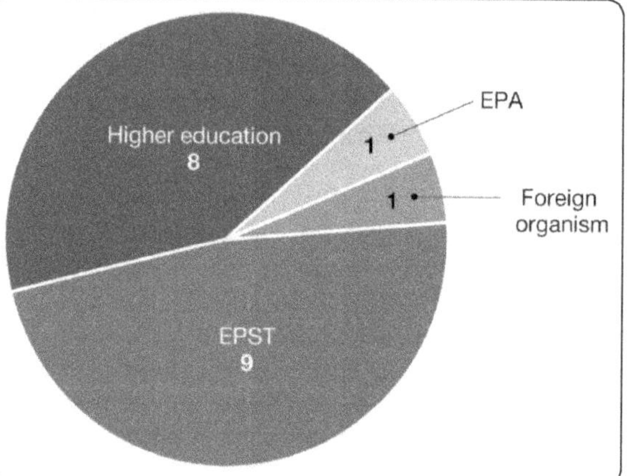

EPST: *établissement public à caractère scientifique et technique* – Public Scientific and Technical Establishment; EPA: *établissement public à caractère administratif* – Public Administrative Institution.

The members of the expert committee hail from 10 research and higher education organisations. Their affiliations demonstrate significant diversity (Figure 1.5): only 26% of experts are affiliated with INRAE. A substantial proportion (41%) are from higher education institutions. Notably, there is only one non-French expert.

Occasional contributors were consulted on specific issues and acted under the responsibility of the expert who requested their input. While they are not members of the scientific expert committee, they are part of the working group and are among the report's authors.

Literature analysed

The complete report cites a bibliographic corpus of nearly 1,800 references. This has been compiled from international bibliographic platforms such as Web of Science (WoS) and Scopus, and has been supplemented where necessary by French-language literature, reports, books, and articles that are not referenced on these platforms.

The targeted knowledge is applicable to the soil and climate contexts in mainland France. Articles were selected with a focus on existing literature reviews when they were relevant to the issue under study (Table 1.1).

For the most part (95%), the cited literature dates from after 1990. Older references were also used to trace changes in stakeholders' perceptions and to draw on certain older methodological sources that remain relevant today. Nevertheless, over half of the corpus was published within the last 10 years (Figure 1.6).

Table 1.1 Types of documents from the corpus cited in the study report.

Type of documents	Number of references	% of the report's corpus
Article	1,500	84%
including literature review	257	14%
Book or book chapter	164	9%
Report	81	5%
Conference proceedings or paper	28	2%
Thesis	13	1%
Other (e.g., dataset, map)	9	1%
TOTAL	1,795	100%
Number of references in at least one database (WoS or Scopus)	1,366	76%
Number of references in WoS	1,278	71%
Number of references in Scopus	1,341	73%

Figure 1.6 Temporal distribution of the corpus cited in the study report (n = 1795)

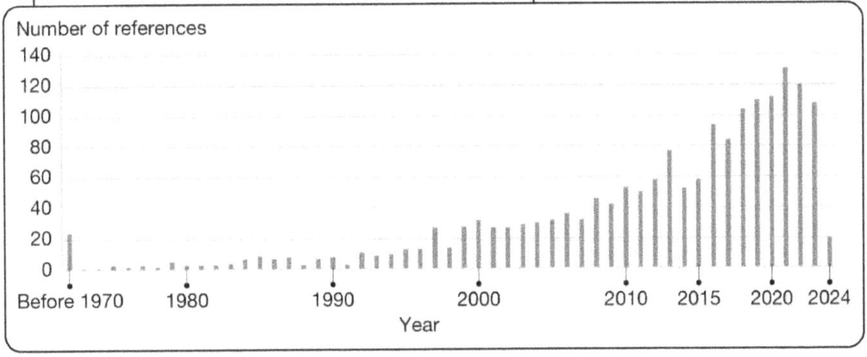

The broad multidisciplinary nature of the corpus is represented in Figure 1.7, which is based on the 1,280 sources referenced in the WoS. The top 15 WoS categories to which the journals publishing the cited articles belong are shown, bearing in mind that the 500 references not included in this WoS category classification mainly relate to grey literature and articles in the humanities and social sciences. The figure also illustrates the core disciplines of the analysed sources in the fields of soil, earth and environmental sciences. In addition to these fields, a broader perspective is provided by ecological studies related to soil functions, and sustainable resource management (water, biodiversity and agriculture; forests, not shown here, appear at a more detailed level).

Figure 1.7. Research areas of the 1,280 references classified in the WoS categories (top 15 categories).

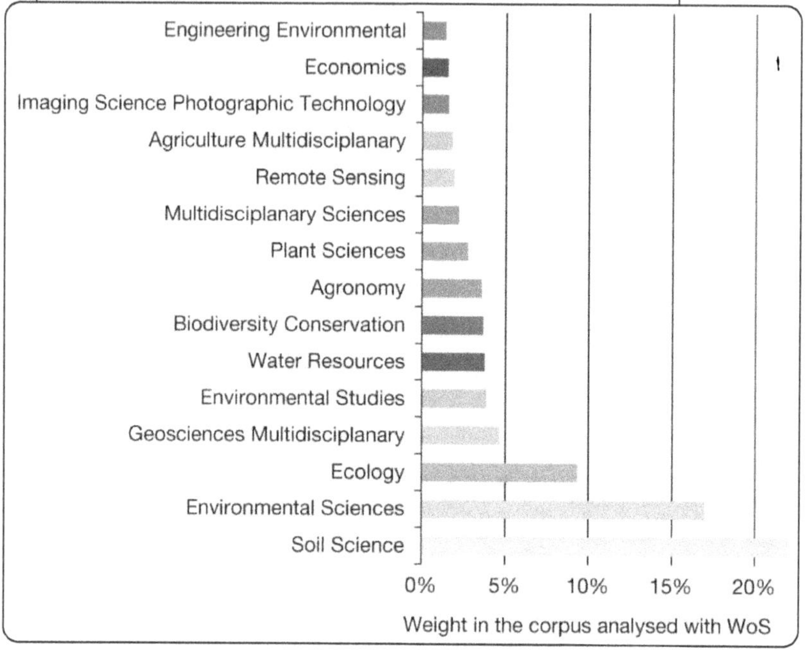

Certain measurement methods based on technologies, identified more specifically by the WoS, are also included such as remote sensing and imaging.

The relative weight of the countries where the authors' affiliated organisations are located (Figure 1.8), reveals the general predominance of the most dynamic geographical areas in terms of scientific production (United States, EU, China). However, it also highlights France's relatively important position, which is linked not only to the selection criteria for sources researched for their applicability in a French context, but also to the historically developed investment in French research on soils. Other countries known for their significant investment in this area are also present in this figure, such as Australia, Canada, Switzerland and Brazil.

Reading guide and terminology used in this condensed report

Reading guide

The bibliographic corpus thus compiled reveals the great heterogeneity of the conceptual frameworks and terminologies used. Soil quality appears to be studied from a variety of angles, including soil condition and properties, threats to its integrity, degradation dynamics, vulnerability, resilience, health, value, safety, ecosystem processes, functions

Figure 1.8. Relative weight of the countries where the affiliations of the authors cited in the report are located, based on the 1,280 references analysed in the WoS (number of references/country > 50).

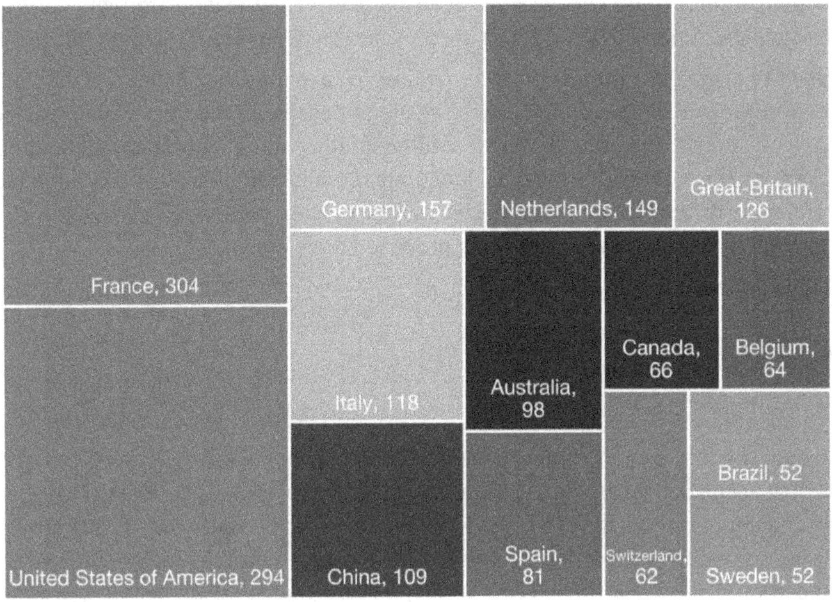

and services (whether realised or potential), with a wide range of associated concepts such as fertility, contamination and pollution, multifunctionality, living soil, nature's contributions (to populations, to nature, as culture), and sustainability of resources.

From a scientific perspective, this heterogeneity poses a challenge, as concepts need to be unified in order to compare, compile, deploy or transfer the produced results. It also presents challenges at the social and political levels, where collective action is only possible when objectives converge. almost importantly, it presents an obstacle at the intersection of these areas, balancing the robustness of scientifically established universal knowledge with the flexibility necessary for stakeholders to share common perceptions and objectives.

This is why this condensed report is structured into two main parts.

The first part observes how soil quality is approached in the social sphere (Chapter 2) and in governance frameworks (Chapter 3). At this stage, the heterogeneity of vocabulary is acknowledged as a fact from which lessons can be learned. This is instructive in that it reveals the foundations on which to build a common language. It demonstrates that such a language is rarely adopted in a strictly top-down manner. A shared vocabulary must be established within the soil quality assessment system, according to its purpose, in conjunction with the relevant stakeholders. With this in mind, scientific resources in the biotechnical fields are mobilised.

These resources are then compiled and organised to support the assessment process. The first step is to outline the various stages of the process and the decisions that must be made at each stage, as these significantly impact the interpretation of the resulting data (Chapter 5). Once these choices have been made explicit, the indicators can be applied appropriately. Around fifty indicators are presented in Chapter 6 to support this process.

These two stages are outlined in Chapter 4, which explains the common ground established in the study, based on the heterogeneity described and the needs revealed in Chapters 2 and 3. The rationale behind differentiating soil quality from soil health and addressing them through ecological functions is explained here. This chapter enables articulation of the lessons learned in the humanities and social sciences with the biotechnical foundations on which the proposed tools are based.

Finally, Chapter 7 presents the conclusions and lessons learned throughout the study.

Terminology used

Soil quality/health

Soil quality encompasses various dimensions, which distinguish it from the concepts of soil quality and soil health. In the biotechnical field, the approach adopted in this study (see Chapter 4) is to consider quality in relation to a specific soil type (e.g. degraded luvisols) or a group of soils within a particular area (e.g. soils within a district). The quality of these soils is assessed based on all their properties, whether inherent (i.e. stable over several decades in the absence of human intervention) or dynamic (i.e. subject to change as a result of use and/or practices). This assessment qualifies a range of possibilities for a soil type, i.e. its potential. The health of a soil in a given location is then assessed by comparing its actual values with the potential values of that soil type or area. In the humanities and social sciences, in the legal context and in everyday language, quality broadly refers to issues relating to soil characteristics and their consequences, without always distinguishing between potential and actual value. For example, when soil is considered for its heritage value or relative rarity, potential and its realisation are not relevant considerations.

To cover all of these topics, this condensed report uses the terminology presented in Figure 1.9. The term "quality" refers broadly to all aspects of soil characteristics and their consequences. This vocabulary is more precise when referring to "performance potential", which is assessed in relation to observed performance in comparable or reference situations. The degree to which this performance is realised, taking into account the uses and practices implemented, is referred to as "health". This distinction between quality and health applies regardless of the dimension measured, whether functions or degradation, for example.

Reference framework

In the biotechnical field, the reference framework for interpreting an indicator includes its definition, measurement method, and reference values against which measured values are compared.[7] However, for a given assessment objective, there are often several possible indicators. Likewise, for a given indicator, there are often several available measurement methods and several possible ways of using reference values. Therefore, the reference framework must be considered in relation to the assessment's objectives and context, taking into account the availability of data. This reasoning encompasses all the elements that constitute the "indication system", as outlined in Chapter 5.

Figure 1.9. Use of the terms "soil quality" and "soil health".

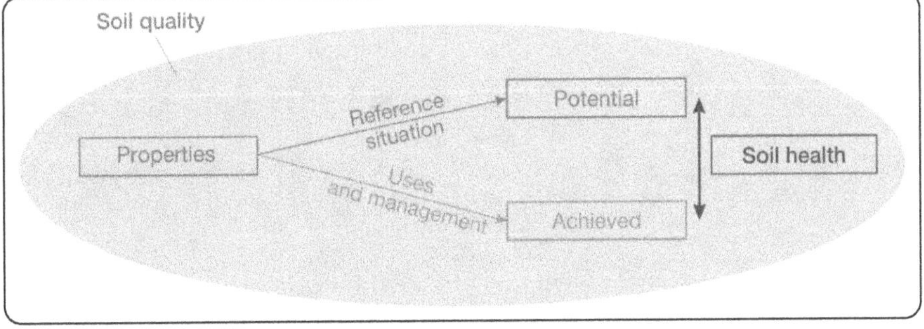

Indicator

An indicator is defined as something used to obtain and transmit information about an object or phenomenon of interest, known as the *indicandum*. It differs from a simple measurement of a parameter in that it has meaning in relation to the object of interest. Therefore, the indicator is considered to necessarily "indicate something", rather than merely being a measured or calculated quantity. The latter can be referred to as a property or parameter. The term "property" is more closely related to soil process dynamics, while "parameter" refers to a quantity that can be used in calculations or modelling. Thus, property, parameter and indicator may in some cases refer to the same quantity. For example, nematode diversity is a biological property of soil, it indicates the fulfilment of the function of "supporting biodiversity" and acts as a parameter in the calculation of certain biodiversity indices.[8] Therefore, it is primarily the use of the information that determines the status of the quantity in question. The indicator is defined by its role in establishing a relationship between a scientifically measured quantity and the perception of the *indicandum* by the actors.

7. See Chapter 5, section "Reference framework and interpretation framework", p. 98.

8. See Chapter 5, section "Relationships between indices, indicators and measured quantities", p. 86.

2. Implementation frameworks for soil quality assessment

Actors and mechanisms involved in soil quality assessment

In order to address soil quality assessment, we propose providing an initial overview of the key stakeholders and mechanisms that are central to its implementation. The table summarised in this chapter starts at the most local level, where the soil manager operates, and then gradually broadens the scope to consider the service providers with whom the manager interacts, up to increasingly collective levels of organisation.

Soil managers

Whether acting individually or collectively (company, association, local authority, state, etc.), the manager is the decision-maker and is responsible for human interventions affecting the soil in question. He may be interested in soil quality for various reasons. These reasons may include personal interest (e.g. acquiring naturalist knowledge), preventing risks associated with land use (e.g. the risk of collapse, or contamination by pollutants), deciding on a land use method (e.g. choosing which crop to plant) or adapting management practices (e.g. using forestry methods to protect the soil from compaction or erosion).

Technical consultants and service providers for enterprises

Technical experts can assist with soil quality assessments. These specialists' access to scientific tools and knowledge that individuals do not, such as measuring instruments that cannot be deployed on an individual scale (e.g. laboratory analyses or the use of proxy-detection/remote sensing tools) and their interpretation reference systems. This advice may be offered by public bodies (e.g. ONF, ADEME, CEREMA, research organisations), consular bodies (e.g. chambers of agriculture) or private service providers. In recent decades, the development of digital tools, combined with an awareness of the importance of soil to the sustainability of living conditions and human activities, has led to a proliferation of service offerings by privately owned companies.

Local authorities' technical experts and urban planning consultancies

Historically, the work of planning offices and local authority technical services primarily has focused on studies verifying the compatibility of soil characteristics with construction or urban development projects, based on geotechnical data. Soil quality is rarely used as a criterion for designating areas for urban development (Serrano and Vianey, 2014). These stakeholders report difficulty in incorporating the diversity of soil parameters and heterogeneity into their considerations. They cite a lack of operational tools for urban planners, landscape architects, and architects that would allow them to "work with the existing soil in order to choose the most suitable crops" (Consalès *et al.*, 2022). For instance, providing ready-to-use soil data in the former Languedoc-Roussillon region "generated numerous requests from public inter-municipal cooperation establishments [...] and private service providers [...] to support the development of territorial diagnoses" (Balestrat *et al.*, 2011), particularly with regard to the biology and ecological dynamics of soils.

Comprehensive public policy frameworks

As decision-makers, land managers are responsible for the activities carried out within their purview, provided they comply with the legal framework and policy objectives that are set at a higher level. This framework comprises various levels of decision-making (from municipalities to European and global bodies) and covers different areas (urban, agricultural, forestry and environmental policies, etc.). It is based on indicators that serve as criteria for defining zoning, establishing rights and obligations, supervising project implementation and granting incentives. Soil quality assessment is required for monitoring purposes, in order to set policy objectives and document the results achieved. To support these monitoring needs, a scientific interest group (GIS) was set up in 2001 at the interface between public policy and research, the GIS Sol. Its mission is to establish and manage an information system that brings together the results of various programmes for characterising and assessing soil quality in France.

Participatory mechanisms

Participatory science and research are defined as "forms of scientific knowledge production in which non-professional scientific actors, whether individuals or groups, participate actively and deliberately" (Houllier *et al.*, 2017), with the aim of solving socio-economic or environmental problems, among other things. Over the past twenty years, such mechanisms have undergone significant development in the field of soils, particularly in agricultural (e.g., AgrInnov project; Ranjard, 2016) and urban (e.g., BISES 2020-2024 project; Ranjard, 2020) contexts. This development has been driven by the need for stakeholder involvement to improve soil management practices. In agriculture, for instance, some studies recommend adapting communication strategies to farmers own perspectives and language regarding soil quality or health.

These initiatives vary greatly in terms of their objectives, multidisciplinary scope, methods and implementation methods. Guidance on sampling and measurement methods is available for initiatives such as the Participatory Earthworm Observatory (OPVT)[9] and the *AgrInnov, ProDij'* and *Clé de sol* projects. All of these schemes are based on interdisciplinary approaches and make use of a wide variety of tools, including videos and forums

9. https://ecobio.univ-rennes.fr/ecobiosoil (accessed on 31/10/2024).

for exchange, such as living labs and workshops. This results in shared learning in terms of conceptual frameworks, knowledge (both theoretical and practical), and levers for action. This positive synergy values academic knowledge and local knowledge equally.

Participants can be involved in a variety of ways, ranging from simple crowdsourcing (contributing to data acquisition) to co-interpretation (participating in the interpretation of results with researchers), co-construction (participating in project development, defining research questions and the approach), or co-responsibility (involvement in logistical, technical and financial support). This strongly determines the scope of the system in terms of producing indicators or methods, acquiring data, and taking actions based on the use of these indicators. Some systems emphasise on an inclusive approach to defining the issue and objectives to be achieved, and the means of responding to them (choice of tools, implementation and interpretation). Others rely on the involvement of groups of citizens or professionals to gain insight into a far greater diversity of situations (soil and climate contexts, land use and management practices) than can be achieved within the framework of conventional experimental approaches. This improves the representativeness of reference databases, particularly for topics that are currently poorly documented (urban areas, forest areas and private properties).

Diversity of perceptions of soils and their qualities

Social factors

Different types of stakeholders have varying perceptions of soil quality and different ways of assessing it, depending on their activities, concerns and social situations. This diversity can lead to misunderstandings and conflicts between stakeholders, so it must be considered when developing a shared approach to soil quality.

At the European level, a study was conducted as part of the LANDMARK project (Bampa *et al.*, 2019; O'Sullivan *et al.*, 2018) to identify the obstacles to implementating a soil management optimisation scenario in a water catchment area, with the aim of achieving the EU's sustainable development goals (O'Sullivan *et al.*, 2018). The study shows that the implementation of the optimisation scenario depends on differences in the prioritisation of soil functions according to the positioning of stakeholders, as well as on the available knowledge (including know-how and technical solutions), institutional constraints (e.g. incentives for implementation), and cultural parameters and informal rules. As an extension of this study, a survey highlighted the differences in the scale at which stakeholders address soil-related issues (Bampa *et al.*, 2019). Regional, national, and European stakeholders tend to consider issues on a large scale, relying on general knowledge and having limited understanding of local contexts. In contrast, farmers and agricultural advisers tend to address them locally, drawing on extensive local knowledge and technical skills, although they acknowledge a lack of technical references (Gascuel-Odoux *et al.*, 2023; Jónsson *et al.*, 2016).

Many studies have sought to identify the factors that determine these differences in perception and concern regarding soil quality. In agriculture, for instance, differences emerge depending on the production model implemented by farmers (e.g. organic farming, conservation agriculture or conventional farming). These differences appear to be partly linked to the material characteristics of the farms (size, technical and economic orientation, etc.) and the farmers themselves (age, training, etc.). However, more recent studies also emphasise the fundamental importance of the social norms and networks of relationships adopted by farmers. Differences in soil management practices among farmers can be political (reflecting the defence of different approaches) and identity-related (reflecting how the farmers describe themselves). A survey conducted in Burgundy (Compagnone and Pribetich, 2017) shows that organic, conservation and conventional farmers have different views on the implementation of no-till farming and that these views are linked to their identity as farmers. Similarly, a study conducted in Burgenland, Austria (Wahlhütter *et al.*, 2016) emphasises the significance of soil in shaping farmers' identities. Farmers distinguish themselves from other farmers, groups, or areas of work based of aspects relating to soil quality or their soil management strategies. In both organic farming and conservation agriculture, these practitioners' relationship with their soil is particularly important in shaping their agricultural identity.

The social embedding of knowledge production

Soil quality cannot be perceived directly or uniformly by all stakeholders. Indicators are therefore essential for creating a shared representation of this quality and making it objective. Scientific knowledge is mobilised for this purpose, and the development of this knowledge is itself embedded in a social context.

Since the 1980s, research conducted in the humanities and social sciences, particularly in the field of science and technology studies (Jasanoff, 2010), has highlighted four main social dimensions that influence the production of scientific knowledge about soils. The political dimension essentially involves political decision-makers prioritising issues and selecting funded research programmes (Bispo and Schnebelen, 2018; Fournil *et al.*, 2018; Hassenteufel, 2010). The cognitive or normative dimension comes into play in the definition of categories and the establishment of thresholds attached to indicators. From a normative perspective, thresholds enable the quality of the soil to be judged as good or bad. These distinctions are based on general knowledge of the situation, which goes beyond the strict framework of the data used by scientists. This requires debate between experts, decision-makers, and practitioners (knowledge of other disciplines, practitioners' techniques, the requirements of different stakeholders for various uses, etc.) (Jónsson *et al.*, 2016). This social embeddedness is also evident in the operational dimension, which depends on the availability of the necessary cognitive and material resources for using indicators (Israel-Jost, 2015). Finally, the performative dimension implies that promoting indicators can shift the perceptions and focus of different stakeholders with regard to soil.

Disciplinary perspectives

Even within the academic field, concepts of soil and soil quality can vary. These differences are partly due to the specific characteristics of each discipline, but also to historical changes in the paradigms underlying the research approach. An important key to understanding the different meanings of soil quality, which emerges from a review of the literature and discussions among experts, is the disciplinary angle. The main features characterising these differences in approach are outlined below.

In law, quality can refer in particular to the status of a natural or legal person, legal capacity, along with the standard legitimately expected of a good or service. In environmental law, the interpretation of quality has evolved from an anthropocentric view of the right to a healthy environment to one enriched by various environmental sciences. In the field of water, for instance, quality can be associated with use (e.g. drinking water or bathing water) or, with various indicators defined according to the type of water body, from an ecological perspective. A similar approach has been developed in relation to soil in French, European and international law. While certain legislative and regulatory texts, as well as judges, refer to the concept of soil quality, the realities covered by this term can vary considerably and are sometimes contradictory.

In economics, the quality of land depends on how it is used, with different uses being linked in terms of value. In this literature, determinants related to land generally take precedence over soil quality in pedological terms. The term "soil" is absent from scientific publications on urban economics and transport economics, where "land" is favoured instead. However, literature focusing on soil properties may consider soil quality to be a cause or consequence of land use choices. This body of work is largely dominated by agricultural studies. Nevertheless, in recent years, there has been a trend towards studies incorporating other land uses, such as forestry and environmental uses (conservation, restoration and compensation), into studies.

Due to the economic challenges associated with developing precision farming tools that utilise large soil properties datasets at a sub-plot scale, economic literature addressing the value of soil quality information is emerging (Tsiboe and Tack, 2022). This value is estimated based on the potential benefits associated with the better decisions enabled by soil quality information.

In the field of sociology, anthropology and geography, soil quality is understood by observing how humans and their social organisations perceive and establish it within systems of norms, and how they integrate it into their decisions regarding soil use and management. While these disciplines do not influence what constitutes good or poor soil quality, they shed light on the mechanisms by which human groups construct and deconstruct these norms in relation to their territories.

The fundamental disciplines of the material and life sciences that relate to soil concentrate on processes closely associated with their respective areas of expertise. Physics plays a major role in understanding soil structure and hydrological dynamics. Chemistry is heavily involved in understanding the transformation of nutrients and contaminants. Biology is essential for understanding the dynamics of living soil communities at all scales and across all domains of life: *Animalia* (animals), *Plantae* (plants), *Fungi* (mushrooms), *Bacteria*, *Archaea*, *Chromista* (chromists), and *Protozoa*. However, these three disciplines are inseparable when it comes to understanding fundamental soil processes such as pedogenesis, plant nutrition, and major hydrobiogeochemical cycles, the dynamics of which combine abiotic and biotic elements within the soil. Some soil functions indicators are therefore difficult to classify according to their physical, chemical or biological nature (organic carbon content, for example). These basic disciplines interact closely with pedology and ecology.

Pedology is the study of soils, involving the establishment of a typology and the explanation of formation and evolutionary mechanisms. Traditionally associated with agronomy, the discipline has tended to prioritise agricultural productivity as an indicator of quality. The chemical component of fertility has also dominated, in line with scientific advances in this field throughout the 20th century. However, advances in soil ecology in recent decades (microbiology, nematology, etc.) have shifted the focus of soil functioning studies toward living organisms.

Ecology considers soil in various ways: as a habitat for organisms; as an ecosystem in itself; or as a component of a larger ecosystem. Similar to how the social sciences study social systems, ecology examines the dynamics of interactions and the evolution of ecological systems. There is no reference framework for judging their quality. Consequently, the ecological quality of environments is often implicitly associated with the quality of life that these environments provide to human beings.

Designations of soil quality

Quality is an ambiguous term in that it can be used descriptively to refer to what the soil is, i.e. its type, and more normatively to judge what the soil is worth, in terms of its quality. This judgement also implies an idea of what the soil should be, which will influence action.

What soil is or what soil does

The designation of "what the soil is" is broken down into two complementary approaches: a typological breakdown and an analytical breakdown.

In a typological approach, soils are classified using vernacular names, which are often associated with colour (e.g. "white soils", "brown forest soils"). These names, can be harmonised in relation to professional and/or land issues (e.g. "ardent sands" in the Indre department, as part of a classification based on agronomic potential), as well as in scientific classifications that are subject to conventions at various levels (French soil reference system,[10] *World Reference Base* [WRB] at the global level),[11] or the frequently used *US Soil Taxonomy*[12] in the United States.

10. https://www.afes.fr/les-sols/referentiel-pedologique (accessed on 31/10/2024)

11. https://www.isric.org/explore/wrb (accessed on 31/10/2024).

12. https://www.nrcs.usda.gov/resources/guides-and-instructions/soil-classification (accessed on 31/10/2024).

The aim of analytical decomposition is to describe the various properties of soil that contribute to its quality, such as its physical, chemical and biological composition (e.g. sand/silt/clay content, nitrogen content, contaminant content, organic matter content and the abundance of organisms), its properties (e.g. conductivity), and the processes that take place within it (e.g. the rate of organic matter mineralisation, enzymatic activities and water or gas transfers).

What soil is worth

Describing soil as "good", "healthy", "rich", "fertile" or "living" amounts to assessing its value. These designations are based on different assessment criteria depending on the category of actors involved, because the classification systems they use to make judgements are specific to their way of thinking and linked to their concerns. Therefore, the literature in the humanities and social sciences encourages us to clarify and contextualise any normative framework used to establish what constitutes good or poor soil quality.

When it comes to land users, the designation of quality is usually linked to the most restrictive parameters, or, on the other hand, the most sensitive ones to their activity. For example, soil can be considered as a surface entity to be managed according to perceived needs in terms of food production, water resource regulation, carbon storage, risks, support for anthropogenic activities (economic, housing), or even habitats/natural spaces for recreational and/or conservation purposes. Earthworkers understand soil in terms of erosion, compaction and dissolution, and must protect it against these pressures. Other stakeholders, such as farmers or foresters using heavy machinery, will face compaction problems, while irrigators will face salinisation, etc. Among farmers, winegrowers have specific characteristics; they understand soil quality not only in terms of its productive potential, but also in terms of its uniqueness as a component of the terroir that shapes the identity of the final product.

What is expected of soil

Due to these differences in conception and perception, conflicts may arise between different types of stakeholders carrying out different activities when assessing the quality of the same soil. The implementation of sustainable soil management is thus hindered by these divergent interests. While farmers and other stakeholders in the value chain prioritise economic criteria when considering soil sustainability, those involved in environmental protection emphasise environmental criteria.

Fournil *et al.* (2018) identify the logic of "environmental requalification" of soil, which has been in operation since the mid-2000s. This differs from the prevailing agricultural perspective, which views soil is seen as merely a fertile medium for production. They distinguish two approaches to environmental requalification: one focuses on conserving soil and its biodiversity *per se* in the face of threats of degradation, while the other emphasises the importance of ensuring the security of the functions and services provided by soil.

Co-production of information on soil quality

Scientific knowledge is therefore one of a number of factors that influence stakeholders' assessment of soil quality. Literature dealing broadly with the use of indicators for public decision-making has conceptualised the interactions between science and decision-making in various models, which are summarised in Box 2.1.

Box 2.1. Main conceptualisations of interactions between science and society (Pülzl and Rametsteiner, 2009).

The transfer model adopts a linear conception of knowledge circulating from science to society. Science is viewed as an area where knowledge is produced, devoid of any value judgements and separate from the world of governance, where knowledge is used. While this conception of interactions is widely questioned in favour of transactional approaches, it remains prevalent in the language habits used by actors to position themselves.

The transactional model considers the formulation of indicators to be a hybrid activity involving interaction between various actors. It takes into account the effect of requests made to scientists on the framing and orientation of research questions, as well as the reinterpretation and adaptation of the results according to their usage context. Both scientists and other social groups are both bearers and producers of knowledge and values.

Boundary spanning (Pülzl and Rametsteiner, 2009) emphasises the importance of continuous translation between the scientific field and the stakeholder field to guarantee successful evaluation. This includes and makes explicit the negotiation processes between the different frames of reference and areas of concern that underpin the positions of scientists, decision-makers, and land users.

In these interaction dynamics, the indicator serves as an intellectual and practical tool. Among the many characteristics that can be used to understand soil, it highlights the ones that are relevant for comprehension or action. At the same time, it draws on and proposes perceptual elements that enable these characteristics to be assessed in concrete terms. Due to its position in interactions between actors, some authors consider the indicator to be a "boundary object" (see Box 2.2).

Thus, constructing an indicator involves cycles of mobilisation and enrolment of private and public actors, including scientists, policymakers, interest groups and citizens. This process is inseparable from a debate on the characterisation of the issues to be addressed, what constitutes the problem and its sub-categories, and the indicators that are supposed to reflect them. Definition issues occupy a significant part of this debate, highlighting the importance of agreeing on a common language.

The selection of indicators by scientific experts is the result of a process of co-evolution. This takes into account stakeholders' concerns, the information received on soil quality (interpretation and sharing), and stakeholders' perception of this quality. Figure 2.1 illustrates the social circulation of information on soil quality and the main modes of interaction

Figure 2.1. Co-production and use of soil quality indicators.

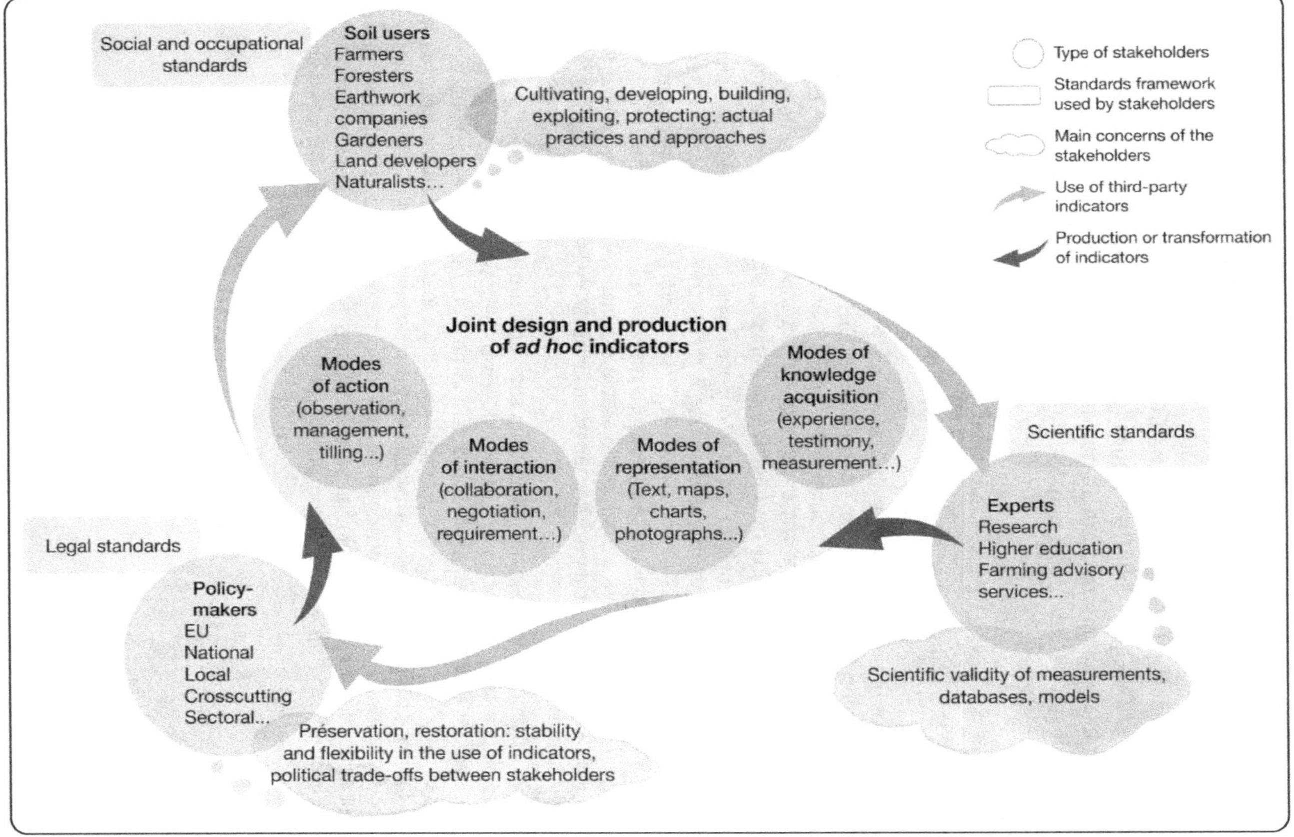

between three broad categories of stakeholders: land users (e.g. farmers, foresters, urban planners, naturalists) technical and scientific experts, and public decision-makers. The normative frameworks to which each group of stakeholders refers are outlined, and the diversity of concerns underlying their positions is illustrated with a few examples.

Many authors therefore advocate producing reference frameworks within the context of action-research projects: "For information to be transformed into knowledge, a variety of stakeholders must collectively question and negotiate the meaning they attribute to it, for example through participatory modelling phases and prospective scenarios, so that it can be used for a purpose co-defined by the various stakeholder groups" (Plant *et al.*, 2021).

Box 2.2. The indicator as a boundary object.

The concept of the "boundary object", which originated in the sociology of science (Star and Griesemer, 1989; Trompette and Vinck, 2009), describes epistemic objects (such as theories, databases and natural history collections) "which are both plastics enough to adapt to local needs and the constraints of the several parties employing them, yet robust enough to maintain a common identity across sites." (Star and Griesemer, 1989).

A boundary object can belong to several social worlds simultaneously, each of which may assign it different identities. However, it facilitates communication between these worlds. Some authors (Turnhout *et al.*, 2007) therefore argue that truly effective indicators are always boundary objects whose meaning changes depending on the actors involved but which have a sufficiently stable structure to be recognised in a shared manner. Consequently, they are primarily vectors of translation and coordination between social worlds.

These authors suggest that indicators that are too finalised and specific may be a weakness if they do not enable stakeholders to negotiate their content and scope of application. Therefore, an indicator that can be reformulated as political or organisational imperatives evolve, enables negotiation and consideration of factors that were not initially formalised. It also allows for diversity of interpretations and appropriations in the social worlds that the indicator connects. These characteristics enabled various ecological water quality indicators in the early 2000s to "facilitate discussion, negotiation and decision-making" between science, policy and other stakeholders (Turnhout *et al.*, 2007).

3. Governance frameworks and soil quality

Private property and public intervention

Land as common, semi-common or trans-property

Some economists view the historical evolution of land status as collective, private or public property as a marker of the major transformation undergone by today's developed economies. In these countries, land is currently largely subject to property rights, whether held by individuals or legal entities such as companies or local authorities. The scope of these rights varies depending on competing rights, such as easements, expropriation or pre-emption under the safer pre-emption right, zoning in urban planning documents and environmental regulations. Nevertheless, they generally grant owners the exclusive right to decide how these areas are occupied, managed and sold.

The fact that land ownership includes ownership of the soil is often cited as the reason why public measures to preserve soil quality have not been widely adopted. A common point of comparison is the legal framework for water or air quality, which are classified as common goods protected as part of the nation's heritage (Jürges and Hansjürgens, 2018). However, throughout the EU, property rights permit a great deal of flexibility in terms of actions that both degrade and preserve the soil.

The cited articles on legal doctrine clearly demonstrate that property rights and the implementation of a soil conservation regime are not incompatible. As part of the environment, soil has constitutional value, just as property rights do. The legislator can therefore balance the competing interests. Ost (1995) refers to the transpropriation mechanism, which exists for other interests such as historic monuments and water quality. This mechanism separates ownership of land from its qualities, meaning the owner of a property is recognised as responsible for, and guarantor of, preserving the soil's qualities.

Beyond the European and French legal framework, the anthropological science observes the existence of a bundle of land tenure rights which considers land ownership independently of the founding categories of Roman-Germanic civil law *(usus, fructus* and *abusus)*,[13] distinguishing between what pertains to soil and what pertains to land.

Under this logic, landowners could retain use of the space, provided they adhere to certain guidelines. Meanwhile, the community of occupants, considered over a wider area, would be responsible for managing natural resources and ensuring the sustainability of soil functions and services.

---

13. *Usus*: right to use the thing (e.g., access); *fructus*: right to enjoy the thing (e.g., cultivate, make fruitful); *abusus*: right to dispose of the thing (e.g., sell).

In these configurations, soil is considered "common" or "semi-common" depending on the author, and is subject to management by a group of actors that may be different from those involved in land management (conceived as a simple surface).

Therefore, property rights do not appear to be incompatible with legislators creating obligations to preserve soil quality. However, the lack of harmonisation of rules at a European level is identified as a major cause of inaction at a national level.

Soil quality governance in view of water and air quality governance

Quality objectives for water and air are set by law and specified by indicators initially designed with human health issues in mind. These have since been supplemented by indicators of the ecological quality of the environment. Furthermore, legal measures have been prescribed to achieve these objectives. Each of these environments is assigned its own concept of quality. In the case of air, "quality" refers to the presence or absence of contaminants such as ammonia gas, ozone and particulates. The concentrations of these substances are compared to threshold values that determine the quality level. However, unlike air,[14] there is no such thing as "pure soil": all soils inherit elements or compounds naturally derived from their original materials. This natural geochemical background varies from one place to another. For instance, serpentine soils naturally contain high concentrations of nickel, chromium, and cobalt, originating from the parent materials rather than human activities.

While the approach based on contaminants was valid for water in the past, this is no longer the case today. Indeed, both law and public policy now incorporate the "terrestrial" dimension of water, encompassing aquatic environments (e.g. riparian vegetation).[15] As Meynier (2020) points out, "the approach to achieving good water body status involves defining scientific quality indicators associated with classification systems based on a reference. Then, a monitoring system is adapted to these indicators, as well as to environmental improvement measures or, at the very least, to preventing environmental degradation. The requirement for water bodies to be in good condition is therefore based on scientific criteria for water quality. These criteria are used to classify water bodies according to their ecological and chemical status, or even quantitative status for groundwater". Furthermore, soils, like aquatic environments (including wetlands), have positive characteristics that contribute to their quality beyond the absence of contaminants. For soils, these include organic matter content, aggregate stability, and biodiversity. However, while it makes sense to apply the concept of water quality to soil quality in order to protect all environments under the law, this would be limited

14. European Commission Communication COM (2013) 918 final, *A Clean Air Programme for Europe*. https://eur-lex.europa.eu/LexUriServ/LexUriServ.do?uri=COM:2013:0918:FIN:EN:PDF. This programme lays the foundations for the first of the thematic strategies announced in the Sixth and Seventh Environment Action Programmes.

15. Riparian forest is an area of exchange, known as an ecotone, between terrestrial and aquatic environments.

by differences in legal status (soil is mostly privately owned, unlike water) and differences in how they function. A very distinct conceptual approach will therefore need to be translated into governance terms.

The NORMASOL project report (Farinetti, 2013) remains the most exhaustive work to date for a comprehensive approach to the links between soil and water quality. Rather than merely referring to uses or services, it refers to "an "ideal" state, which is the undisturbed state, to which we must strive to approximate as closely as possible" (Farinetti, 2013).

Doctrine of public intervention on land

Decision-making processes concerning soil are broadly defined by Juerges and Hansjürgens (2018) as "the sum of all formal and informal institutions (e.g., legal requirements, the legal framework, market incentives, rules, standards, habits, attitudes) that affect the land-related decision-making processes of state and non-state actors at all levels of decision-making."

With regard to the public component, i.e. the legal and institutional aspects of land governance, the question of the most appropriate balance between the collectively established framework (and its territorial scale) and the autonomy granted to private actors in decision-making is often raised. In its most basic form, public economics considers private land use decisions to be effective when the consequences of these decisions are fully internalised, i.e. when they result in a gain or loss for the decision-maker. In the presence of externalities, i.e. consequences that are not taken into account by the market for third parties, public intervention is necessary to avoid or compensate for these consequences.

Authors who assume that there are few or no externalities demonstrate that, particularly in the United States, most of the benefits of soil conservation are private. Farmers who prioritise soil quality enjoy a positive return on investment, and improvements to the soil resulting from their practices are reflected in land prices. From this perspective, therefore, the degree of soil degradation corresponds to a socially accepted choice, as has been demonstrated in the case of erosion (McConnell, 1983), for example.

These results are qualified by certain studies in North America which highlight that observed practices depend on soil condition only in cases of high degradation, even though earlier intervention would have been economically advantageous for the farmers concerned. These findings raise the question of why farmers do not strive to achieve optimal soil quality when they would benefit from doing so (Stevens, 2018). With this in mind, a synthesis of 87 European studies cites differences in attitudes towards environmental preservation, skills, farm size, and experience as explanations for this paradox (Bartkowski and Bartke, 2018). Meanwhile, demographic factors such as age, gender, education, and perceptions of neighbours or society are said to have more uncertain effects.[16]

16. See Chapter 2, section "Social factors", p. 25.

The identification of externalities linked to land use is established in spatial and temporal terms. Spatially identified externalities are the consequences, for other areas, of practices implemented on a plot of land. For instance, inputs may be transferred partially to aquatic environments, resulting in a deterioration in water quality that the community will bear; alternatively, the destruction of vegetation cover can affect water infiltration dynamics and lead to flooding downstream of the watershed. This can also lead to increased water erosion, mud flows, and degradation of water quality downstream of the plot. Externalities identified in terms of time correspond to consequences suffered by future users of the plot. For instance, systematic ploughing may increase short-term yields, making it more profitable than the monetary cost of the intervention. However, it has delayed consequences for the soil's structure and organic matter content, which affect this same yield. The consequences of a land user's decisions are then borne by their successor.

For economists, identifying these externalities enables them to determine relevant corrective measures and their respective contributors and beneficiaries. However, due to the difficulty of measuring individuals' and society's preferences for the functions and services provided by soil, these principles have not yet been well established in the literature.

Public intervention measures concerning land

Soil governance involves organising discussions and arbitrations to determine the relative importance of different soil functions and associated use changes. Some authors propose classifying the available instruments, whether public or private, into five categories: legal, planning, market-based, informational, and cooperative (Jürges *et al.*, 2018).

Others (Prager *et al.*, 2011) categorise them into three types of measure: mandatory, voluntary, and informative. With regard to public intervention specifically, which is common to both law and economics, the doctrine is broadly divided into two main types of intervention.

The first type, which is prevalent in administrative law, involves establishing a preventive mechanism of special administrative policing based on the recognition that preserving soil health is in the public interest. This approach has the advantage of intervening upstream of soil degradation by imposing certain rules on activities and creating a system of sanctions in the event of non-compliance. For example, this approach is emerging in urban planning law, with the prospect of establishing a system in which administrative decisions relating to land use and occupation would take soil health into account. However, economists question the effectiveness of such a system due to its rigidity, potential administrative burden, and the information requirements involved in setting means- and/or result-based obligations relevant to society and monitoring compliance with these obligations.

Secondly, the coercive nature of environmental standards is not necessarily a criterion for their effectiveness or adoption by the relevant parties. Instead, voluntary approaches are favoured, such as contractualisation, certification and market policies.

Examples of this can be seen in soil conservation policy in the United States and in the field of carbon storage on agricultural land. Economists favour this approach in order to focus on the functions and services provided by soils that are impacted by private decisions. This involves offering remuneration for practices that are conducive to providing these services, and/or taxing practices that are detrimental. This is currently the most widely used approach for agricultural purposes (Jeffery and Verheijen, 2020), with incentives being offered for practices such as crop rotation, reduced tillage, the use of organic amendments such as compost and manure, as well as intermediate crops. However, a major weakness of this type of intervention is that its effects will only be sustainable if the funding and incentives offered are sustainable. Moreover, the effective provision of ecosystem services cannot be verified by field measurements because only the implementation of practices is subject to controls. Contractualising the provision of ecosystem services ultimately does not resolve the question of whether interventions should be based on obligations of means or results.[17] The technical justification for the granted incentives is the quantification of services provided as a result of implemented practices, based on ad hoc experiments. These assessments are then applied to all implementation situations without verifying the validity of this transfer.

Nature of incentives and obligations introduced

The literature emphasises the importance of basing incentives on obligations of means, such as uses and practices (Jeffery and Verheijen, 2020), provided that the scientific basis on which they are defined is improved and updated. Modelling is promoted in this regard (Bartkowski *et al.*, 2021) as this field of research is still in its infancy. The real effect of incentives on soil quality on a global and long-term scale remains to be documented in relation to the broader issue of evaluating public policies, including windfall effects, inequalities between users, and the resources invested in restoring degraded land compared to measures to prevent such degradation.

Most soil quality governance mechanisms are based on obligations of means and strong assumptions about the relationships between land use, management practices and soil quality, and the resulting benefits in terms of ecosystem services. Indeed, regulatory frameworks are most often established for the sake of preserving these services. For instance, the supply of biomass for food and non-food purposes has resulted in the protection of agricultural and forest land, while the provision of high-quality water has prompted the protection of catchment areas. Similarly, climate regulation has given rise to incentives for storing carbon in soils. However, the measures taken are based on obligations of means, which are deemed to be conducive to the provision of these services without the results always being measured. Furthermore, while emphasising the importance of linking soil quality and services, some authors warn that this approach may "partly obscure the environmental component that produces these services" (Langlais, 2015) by prioritising or even maximising service production.

17. See above, section "Nature of incentives and obligations introduced".

In order to counterbalance the tendency to prioritise the most quantifiable ecosystem services, which can result in the neglect of less apparent aspects, discussions are focusing on an approach to preserving soil quality that considers the heritage value of certain soils due to their rarity or location. According to Article L. 110-1 of the Environmental Code, soils "contribute" to the nation's common heritage. This sometimes leads to their preservation, for example when they constitute the habitat of protected biodiversity.[18] However, the focus on biodiversity can perpetuate the biases associated with the methods used to compile lists of protected species, which are well recognised in scientific spheres. These lists particularly overrepresent bird[19] and mammal species, leaving little room for the protection of invertebrates, despite their strong link to the soil. This approach undoubtedly affects the types of soil that benefit from protection as habitats. Extending this logic, the same question could arise for microorganisms, given our growing knowledge of the specificity of microbiota and their significant ecological role.

In urban areas, planning mechanisms have evolved over the past decade with regard to biodiversity, particularly with regard to preserving nature in cities without necessarily associating it with heritage. The relative scarcity of unsealed urban land offers potential for the implementation of preservation measures, although the relevant legislation is still in development. Local authorities and design offices are gradually incorporating the functional aspects of unsealed land into their considerations, taking into account the services it provides. The concept of "brown corridor", while not yet defined or legally imposed, is being considered, with certain municipalities, including Poitiers, Tours, Paris and Ris-Orangis, exploring the restoration and preservation of ecological continuity based on soil biodiversity.

Measures and economic values of soil quality

Primacy of land over soil

In the economics corpus

The economic literature on soil quality seems to be divided. Most of it (90%) refers to land without distinguishing between different types, focusing primarily on characteristics such as location, legal status and surrounding resources (e.g. water and energy), as well as economic and social interactions between users and uses (e.g. proximity to urban centres and processing and marketing structures for agriculture). Urban and transport economics only marginally address the "soil" component of land, with a few exceptions concerning its stability. In terms of the aspects considered, variations in the biological, chemical or physical characteristics of soils are generally considered secondary in

18. Annex I to Council Directive 92/43/EEC of 21 May 1992 on the conservation of natural habitats and of wild fauna and flora.

19. It should be noted that almost all birds benefit from this protection (Lévy-Bruhl, 1992).

economic trade-offs regarding land use choices in urban contexts. The remaining literature (10%), which explicitly deals with soil, focuses on agricultural uses and, more recently, takes other environmental considerations into account, such as pollution, conservation and restoration. Nevertheless, the main findings remain fairly similar: land use decisions depend more on human parameters (e.g. agricultural product prices, discount rates, and returns on investment) than on biological, chemical, and physical parameter.

In trade-offs between land uses in France

This primacy of land over soil is also documented in social science research, which shows that knowledge of soil quality has little influence on land use decisions. For instance, an assessment of the agronomic potential of the land was compared with urbanisation dynamics in the Calvados department. The research concluded that "good and very good agricultural land has been proportionally more affected by urban sprawl than land of lesser quality" (Gouée *et al.*, 2010). More generally, various obstacles to considering soil characteristics have been identified.

Urban planning documents divide space into sectors (classified as urban [U], intended for urban development [AU], agricultural [A] or natural and forest [N]). This produces framing effects that mask the pedological characteristics of soils and their integration into urban development projects in particular (Consalès *et al.*, 2022). The same is true of the cadastral network, which allocates usage, management and ownership rights according to criteria that do not necessarily consider soil.

Furthermore, regulatory measures that require urban development zones to be located in close proximity to existing urban areas restrict the conversion of lower-quality land within the territory in question. In Rungis, for example, "the distribution of land use between construction and agriculture does not seem to have considered the quality of the soil, despite efforts by producers to identify the 'best land' and measurements of the soil's physicochemical quality." (Gauthier, 2020).

Finally, even when soil criteria are used to distinguish between protected and unprotected areas, they are often used as a secondary criterion after an initial division of space based on land criteria. For instance, in Montlouis-sur-Loire, a municipality within the Vouvray *appellation d'origine contrôlée* (AOC) wine region, the delineation of a protected agricultural area (ZAP) was based on the spatial footprint of the appellation's vineyards (Serrano and Vianey, 2014). Experts have confirmed that such cases of requests for AOC classification as a bulwark against land conversion are not anecdotal.

In light of these findings, researchers have explored ways to reintroduce soil quality into land value calculations. Tafani and Jouve (2023) propose a Potential Buildability Coefficient (PBC), which modulates land value based on the agronomic value of the soil and the timing of municipal construction projects. The PBC increases with the immediacy of the planned building work and is then multiplied by the agronomic value[20] to determine

---

20. The reference framework remains to be chosen. See below, "Agronomic potential: between use, management methods and soil quality", p. 48.

the price per hectare of land. However, this proposal has had little impact. The tool is difficult to use because it defines values that are disconnected from the market value of land, which confuses users and decision-makers. However, the idea of establishing a new value grid that takes into account the multiple functions of land will be retained in order to inform public decision-making.

Value of rural leases and improvement/degradation of land

At the intersection of economics and law, the value of rural leases is a heavily regulated economic parameter under French law. Furthermore, when a lease is terminated, either party may claim compensation based on improvements or deterioration to the land. In the case of improvements, the outgoing lessee receives the compensation, while in the case of deterioration, the lessor receives it.

At the departmental level, land quality is considered when establishing its rental value in accordance with French rural lease rules (Party *et al.*, 2014). Since the 1970s, Indre-et-Loire has used soil data to develop a classification and rating system for agricultural land. Currently, prefectural decrees governing agricultural rents incorporate soil criteria to varying degrees. Distinctions are made firstly according to the type of agricultural land use (e.g. fields, meadows, permanent crops, specialised crops or crops specific to a department or area), and secondly according to sub-departmental areas, partly reflecting the network of "small agricultural regions".

There are two assessment methods: one is based on taxonomic soil types, which may vary depending on the department; the other uses score-based rating scales. The latter method prioritises soil criteria (especially for fields and meadows), which mainly refer to physical characteristics (e.g. depth, composition and stoniness) or water-related characteristics (e.g. hydromorphism and sensitivity to rainfall hazards). These methods are updated infrequently, which partly explains why the criteria are outdated and why chemical and biological parameters are absent.

Disputes concerning the improvement or deterioration of land are based on Article L. 411-71 of the Rural Code, which outlines the methods for calculating compensation. Improvement is deemed to have occurred when the land's production potential increases by at least 20%. However, these legislative provisions are specified in an inventory plan established by the October 31st, 1978 decree, which has never been updated and whose content reflects an outdated conception of land improvement that is partly contradictory to agroecological practices. Consequently, certain case law considers the ploughing up of grassland and the application of mineral fertilisers or plant protection products to be improvements. This approach is based on the perception that soil quality is defined by the yield obtained per hectare. However, establishing a direct relationship between soil quality and productivity per hectare is problematic, as yield depends not only on soil quality, but also on land use (e.g. destruction of hedges, terracing and drainage) and management practices (e.g. fertiliser and pesticide use, and mechanical interventions). While these practices can compensate for quality issues and temporarily mask them (e.g. liming acidic soil, which will need to be repeated), they can also affect the soil's ability to maintain its quality on its own (e.g. mineral fertilisation ultimately inhibits mineralisation capacity). In addition, soil is sometimes considered degraded if management methods do not correspond to standard farming practices and associated yields, which are traditionally used as a reference point. The lessee's duty to farm–"active enjoyment" being opposed to "idle enjoyment" of leased land–is an unspoken rule still widely upheld in rural law and applied by judges. Returning land to its "natural state", i.e. a condition below that which the leased land is capable of achieving, is interpreted as degradation of the property. Therefore, it is these historical practices that constitute the common benchmark against which yields are compared. This does not take into account the specific input-saving practices that some farmers may have adopted to improve the ecological quality of their land while reducing productivity.

Soil quality in economic analysis

Economic analysis examines soil quality in relation to land use decisions, both upstream and downstream. In the former, quality is a production factor (input) that enables the provision of different levels of goods and services. This quality is defined in relation to the expected goods and services and the established uses for achieving them. For example, soil that is of good quality for producing cereals may not be suitable for producing wine. In the second instance, soil quality is an output resulting from decisions relating to its use. Indeed, soil use often leads to environmental changes that can improve or degrade its quality, such as mineral fertilisation or ploughing, which impact organic matter content. Comprehensive economic analyses must consider soil quality in both these ways to take the entire system into account. Furthermore, changes to the soil do not necessarily impact the different aspects of soil quality equally. For instance, draining soil improves its agronomic quality, but reduces its quality with regard to water regulation.

In light of these principles, the status of parameters related to soil quality may vary (input, output or both) depending on the adopted perspective and the issues addressed by the study. In the analysed international literature, fertility, organic matter, available water capacity and bulk density are mainly considered as inputs. This reflects the predominance of economic studies that incorporate soil quality measures, which focus more on the economic repercussions of good or poor soil conditions for agriculture than on the consequences of land use and management choices for these properties.

Economic value of soil

The lack of information in private and public decision-making on soil quality and its evolution has led to the development of economic assessments. The aim of these assessments is twofold: they attempt to place an economic value on soil as an asset and to estimate the potential costs of its degradation. These quantitative approaches form part of a broader effort to evaluate the economic value of natural capital (including soil) and the ecosystem services it provides. The value of these economic indicators lies primarily in their instrumental nature to support users' and public policymakers' decision-making processes, rather than organising the purchase or sale of soil quality as prices might do. Derived from methodologies and data of varying reliability, these indicators provide comparability with other economic decisions and contrast with assessments based exclusively on biological, chemical or physical factors (Rossiter, 1995). The degradation of soil quality can be more directly compared with rehabilitation costs when expressed in monetary terms than when expressed in terms of nutrient quantity or organic matter content.

The usual limitations of monetisation, whether philosophical or practical, also apply to the specific context of soil quality. These limitations concern:

- **The comparability** that reference to monetary value allows. This is both the main advantage of monetisation and the main target of its critics. Monetisation expresses the consequences of choices that affect various entities and actors in a single unit. However, these entities may be considered incomparable or non-substitutable (Baveye *et al.*, 2016), as in the case of the overall monetary balance of an increase in agricultural yield versus a loss of biodiversity;
- **The temporality,** which involves a degree of uncertainty. Soil quality management often requires trade-offs between different time horizons. For example, avoiding the short-term cost of soil preservation shifts the costs of repairing it to the future. As the future is unknown at the time of decision-making, assessments are based on assumptions;
- **The completeness,** which requires consideration of the entire soil system. Monetisation or decision-making rules based on economic criteria require consideration of all the processes involved. This necessitates more detailed classifications of ecosystem services and their links with soil parameters (Dominati *et al.*, 2016);
- **The data** available from soil science research. The spatial and temporal scales and resolutions commonly used in soil science research do not necessarily correspond to those that would be relevant for monetisation or decision support in an economic context (Robinson *et al.*, 2014).

Thus, there is currently no consensus on the economic importance of considering soil quality in the context of this study, particularly with regard to the costs of preserving it. In light of these challenges, economic research has evolved to address each dimension more precisely, even if this necessitates omitting their full articulation. This explains the segmentation observed between agricultural, environmental and urban economics, each of which studies different dimensions of soil quality without necessarily explaining the interactions between them. This restricts the effectiveness of indicators for decision-making, as certain trade-offs between the different dimensions of soil quality cannot be made and not all the impacts of decisions are considered.

Given the difficulty of assessing soil quality in economic terms in its entirety, analyses tend to be limited to the elements believed to be most affected by the consequences of the decisions to be made. In order to avoid the arbitrary aggregation rules required by total economic cost or multifunctional indicator approaches, the focus of this literature is on specific decisions and objectives. While this may mean sacrificing exhaustiveness, researchers are encouraged to act as "plumbers" (Duflo, 2017) and use indicators that focus on the details of decisions which often make the difference. This work has a strong operational component and can be conducted on the basis of real-life experiments, with applications at the individual or collective level (e.g. public policy impact studies). Currently, however, it is rarely applied to land and soils. These approaches complement holistic approaches, which, despite being difficult to quantify, remain important for identifying neglected dimensions. For instance, certain low-impact or long-term effects could be overlooked if the focus is too heavily placed on the most impacted elements. A more comprehensive approach is also necessary for negotiations between stakeholders on what is and isn't comparable, based on the information produced and shared.

Criteria used in the field of law

Unlike other areas of the environment, such as water or air, there is no legislation in the EU or France dedicated to soil conservation. However, references to soil quality are common in many areas of law, particularly urban planning, agriculture, and environmental legislation. These references appear with various and sometimes conflicting meanings, and the fragmentation of provisions relating to soil conservation is widely recognised as a source of inefficiency (Billet, 2016a; Grimonprez, 2019; Hermon, 2018). In France's various codes, the concept of soil quality is often included in lists of elements whose quality must be preserved. For instance, Article L. 101-2 of the Urban Planning Code sets out general objectives and states that local authorities must protect natural environments and landscapes, as well as ensuring the preservation of air, water, soil and subsoil quality, natural resources, biodiversity, ecosystems and green spaces. It also mentions the creation, preservation and restoration of ecological continuity. This remains the case in environmental criminal law, which aims to punish "substantial damage to the quality of the air, soil or water".[21]

Examining the information available at the most centralised levels of law (EU and France) reveals that the soil quality preservation indicators focus more on changes in land use and soil management practices (obligations of means) than on soil quality itself (obligations of result). These indicators are therefore largely based on the assumption that certain types of use or practices are more conducive to achieving the desired outcome than others. At a more decentralised level of governance, the legal framework makes more reference to soil quality indicators.[22] However, effectively using this type of information has depended on case studies being documented in the scientific literature. A more systematic analysis of the indicators used in public policy would require surveys to include these local levels of governance.

21. Art. L. 173-3 C. env.

22. See above, section "Value of rural leases and improvement/deterioration of land", p. 40.

Legal criteria relating to types of use

Measures to preserve soil quality, formulated in terms of land use distribution, primarily focus on limiting the consumption of Natural, Agricultural and Forest Areas (ENAF). In urban planning terms, the inability to build on land is the main factor associated with preserving soil quality. For monitoring purposes, urban planning documents have been required to present an assessment of agricultural land consumption over the previous 10 years and set quantified and justified targets for frugal land consumption since the Grenelle 2 law of 2010. The Urban Planning Code also stipulates stricter provisions for coastal and mountain areas, and gives departments the power to designate, in consultation with the relevant municipalities, Protection Perimeters for Natural and Agricultural Areas in Peri-urban Areas (PEAN)[23] or Sensitive Natural Areas (ENS).[24] Environmental assessments of plans and programmes increasingly (though not yet fully) consider the impact on soil quality of the areas of ENAF consumed or compensated for, largely thanks to developments in the European framework.

The Climate and Resilience Law, adopted in 2021 and supplemented in 2023, gave this objective of restraining land consumption a slightly different and more coercive form. It sets the goal of achieving "zero net land take" by 2050, with an intermediate target of halving ENAF consumption by 2031 (see Figure 3.1). This process is monitored by the National Observatory on Land Conversion.

This law introduces two major changes with regard to monitoring ENAF consumption. Firstly, it replaces the metric of ENAF consumed with that of net artificialisation. Secondly, it imposes a territorial-level limit on net artificialisation, with the relevant territories being discussed and decided upon at a regional level. The main novelty of this approach is the ability to deduct renaturalised areas from the total artificialised area to obtain the "net" artificialisation. The definition of artificialised areas and the measurement of their surface are based on a table cross-referencing occupation and use[25] provided by the Large-Scale Land Cover Observatory (OCSGE), which is based on aerial images. This nomenclature classifies as "artificial" those occupations and uses known to have a lasting impact on ecological functions, but the criteria used to assess this impact are not specified. There are also certain ambiguities and contradictions, such as the status of artificially constructed soils on slabs. Ultimately, therefore, it will be necessary to focus on the functions themselves. Studies and institutional reports suggest that land planning document give a qualitative definition of what is called "*pleine terre*", meaning the surface of open ground required at the scale of a developing project.

Multiple uses raise other issues, and the decree of 29 December 2023 has, for example, regulated the possibility of considering ground-mounted photovoltaic energy production

23. Art. L. 113-15 to 28 of the Urban Planning Code.

24. Art. L. 113-15 to 28 of the Urban Planning Code and Art. L. 113-8 to 14 of the Urban Planning Code.

25. https://artificialisation.developpement-durable.gouv.fr/sites/artificialisation/files/fichiers/2022/05/2022_05_03_Tableau-OCSGE-CouvUsage-ARTIFICIALISATION%5B1%5D.pdf (accessed on 9/11/2024).

Figure 3.1. Regulation of land artificialisation dynamics included in the 2021 Climate and Resilience Law*.

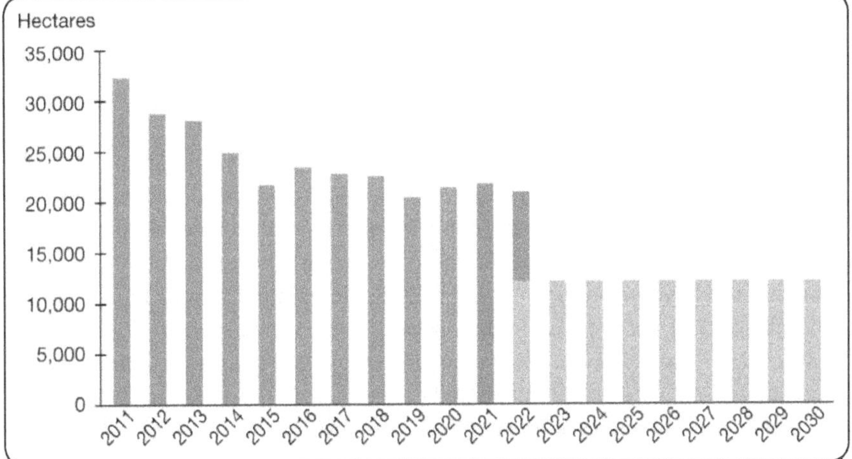

In red: reference period; in green: projected consumption; in blue: annual consumption of ENAF (in hectares), between 1 January 2011 and 1 January 2023**.

*https://artificialisation.developpement-durable.gouv.fr/sites/artificialisation/files/fichiers/2024/07/ZAN%20DP%2027nov23_VF-17_0.pdf (accessed on 8/11/2024).

**https://artificialisation.developpement-durable.gouv.fr/parution-des-donnees-consommation-despaces-2009-2023 (accessed on 8/11/2024).

facilities as non-artificialised spaces (Climate and Resilience Law, Art. 194). It stipulates that installations must be reversible and allow the maintenance of vegetation cover corresponding to the nature of the soil, as well as pre-existing natural habitats where applicable, and significant agricultural or pastoral activity.

Similarly, most other EU Member States are developing quantitative protection measures for areas used for agricultural and/or forestry purposes, particularly with a view to ensuring food sovereignty. This is in line with the "No Net Land Take" objectives identified by the European Commission in 2011[26] despite the absence of a harmonised approach to soil quality under European law. This is particularly evident in France, Italy, the Netherlands, Spain, Belgium and the Czech Republic.

In climate-related matters, soil is considered in greenhouse gas emission inventories based on land use, land use change and forestry (LULUCF).[27] Two approaches are combined, one that accounts for changes in land use categories and one that monitors

26. Roadmap to a Resource-Efficient Europe (COM [2011] 571).

27. Decision No. 529/2013 of 21 May 2013 on accounting rules for greenhouse gas emissions and removals resulting from activities related to land use, land use change and forestry, and information on actions related to these activities; Regulation No. 2018/841 of 30 May 2018 on the inclusion of greenhouse gas emissions and removals from land use, land use change and forestry in the 2030 climate and energy framework (LULUCF).

the net absorption balance by land and forests, with the aim of strengthening long-term carbon sinks. However, achieving specific results in terms of soil carbon content is not required as part of this objective.

Legal criteria relating to management practices

Provisions relating to soil quality preservation are based on criteria targeting management practices that are considered beneficial to soil quality. In particular, they are implemented within the CAP framework, where many measures concern soil more or less directly. These measures can be found in eco-schemes provided for in the National Strategic Plan (NSP), Agri-Environmental and Climate Measures (AECM), or in Good Agricultural and Environmental Conditions (GAEC). Regarding the indicators considered in this framework to be conducive to "monitoring the challenge of improving soil quality", it is stated that "the NSP will measure the proportion of utilised agricultural area (UAA) covered by soil management commitments annually [...] The target to be achieved by the end of the programming period is 74.07% of UAA".[28]

Similarly, financial incentives to store carbon in soils have recently been introduced through market mechanisms. These also focus on the implementation of practices rather than the achievement of targets. For instance, the Low-Carbon label (LBC) was introduced by decree on November 28, 2018.[29] This paves the way for individuals who are not legally required to limit their greenhouse gas emissions to voluntarily offset them (known as the voluntary offset market) by funding third-party projects that limit emissions or store carbon. This labelling is based on methodologies that outline favourable practices and provide a basis for calculating the amount of carbon stored as a result of their implementation. However, as with LULUCF, the actual soil carbon stock is not directly measured. "There is a consensus that it is not possible to determine the change in soil organic carbon stock on a plot in the short term (3 to 5 years) after the introduction of carbon-storing practices through field measurements. Studies have shown that, even with substantial sampling efforts involving 10 samples per plot, it takes approximately 24 years after the practice is implemented to detect a change in carbon stocks. In other words, due to the uncertainties associated with sampling and soil testing, and the short duration of LBC projects (five years), it is impossible, with an acceptable sampling effort, to measure the effect of the project by comparing C stock values measured in situ at the beginning and end of the project." (Soenen *et al.*, 2021).

This voluntary market regulation and these methodologies may be subject to change. On November 30, 2022, the European Commission submitted a proposal for a regulation

28. Plan Stratégique National, p. 306 and following, operational procedures specified by Décret 2022-1755 of 30 déc. 22 relatif aux aides du plan stratégique national de la politique agricole commune, Arrêté du 14 mars 2023 relatif aux règles de bonnes conditions agricoles et environnementales (BCAE).

29. Décret n° 2018-1043 du 28 nov. 2018 créant un label « Bas-Carbone », Arrêté du 28 nov. 2018 définissant le référentiel du label « Bas-Carbone ». Millard J.-B., Bosse-Platière H. (dir.), 2022. Le CO_2 vert capturé par le droit. Le carbone en agriculture et en sylviculture, LexisNexis.

establishing a Union certification framework for carbon removals,[30] with the aim of harmonising the various existing arrangements for the "voluntary carbon market" within the EU and ensuring reliable carbon accounting through certification. While it is not yet clear which methodologies these certifications will be based on, it should be noted that two requirements apply to carbon: "A carbon removal activity shall provide a net carbon removal benefit" (Art. 4) and "aims at ensuring the long-term storage of carbon" (Art. 6). Additionally, the carbon component must be complemented by a guarantee of a neutral or beneficial impact on the following sustainability objectives: "(a) climate change mitigation beyond the net carbon absorption benefit referred to in Article 4 (1); (b) climate change adaptation; (c) sustainable use and protection of water and marine resources; (d) transition to a circular economy; (e) pollution prevention and control; (f) protection and restoration of biodiversity and ecosystems" (Art. 7).

Finally, obligations relating to land management practices are also set out in relation to renewable energy:

• in order to protect certain soils from indiscriminate crop cultivation for biofuels and bioliquids, Directives 2009/28/EC and 2009/30/EC of April 23, 2009,[31] supplemented on September 9, 2015,[32] introduced production "sustainability" criteria. This means that certain soils cannot be used or converted to produce raw materials for biofuels and bioliquids. This applies to: "land of high biodiversity value", "land with significant carbon stocks" and "land with peatland characteristics";

• the Renewable Energy and Biodiversity Observatory, created by Decree No. 2024-315 and implemented jointly by the OFB and Ademe, is responsible for "summarising the knowledge available through existing studies and data on the impact of terrestrial renewable energies on biodiversity, soils and landscapes".[33]

Legal criteria relating to soil quality

Of the criteria included in legally accessible, centrally available texts, few refer directly to soil properties, particularly in relation to urban environments. Such parameters are more commonly found in provisions adopted at sub-national level.[34] An exception is the management of polluted sites and soils, which has a set of standards relating to soil quality. However, the associated thresholds and provisions mainly address human health issues and apply to only a small part of the territory. Taking an approach that focuses on the ecological functionality of soils would require an index that reflects the overall pollution load to which soil organisms are exposed. However, such a metric is not currently available.

30. COM (2022)672 final.

31. Directives 2009/28/EC and 2009/30/EC of 23 April 2009 (Articles 17 and 7b respectively).

32. Directive 2015/1513 of 9 September 2015.

33. https://www.legifrance.gouv.fr/jorf/id/JORFTEXT000049375494.

34. See, for example, the criteria governing the value of rural leases at departmental level, above, section Value of rural leases and improvement/degradation of land.

However, some measures focus specifically on soil types and morphology. These include the delineation of zones such as wetlands and disadvantaged areas, which are defined by criteria including drainage, roc fragments, sand, heavy clays, organic matter, vertic soils, rooting depth, outcrops, salinity, sodicity, acidity and slope,[35] AOC areas are also included.

Agronomic potential: between use, management methods and soil quality

The legal definition of agronomic potential, which is used to delineate permitted uses, takes into account both management methods and soil quality. In urban planning law, agronomic potential is a concept used to identify areas suitable for agriculture, which are classified as Zone A in urban planning documents (PLU or PLUi). Article R. 151-22 of the Urban Planning Code defines these Zone A areas as those that are to be protected due to their "agronomic, biological or economic potential". Of these three criteria, however, agronomic potential is the most widely used. Therefore, using the agronomic potential criterion for land use planning purposes could be considered a crossroads between surface area perspective (deciding land allocation), the consideration of management methods, and indicators documenting agronomic potential.

Departmental Agricultural Land Maps (CDTA) as an attempt to establish a reference framework

Under the 1980 Agricultural Orientation Act, the CDTA were tasked with classifying agricultural land into six categories of "potential productivity" to inform decision-makers of land improvement opportunities and protect agricultural land. However, only 132 of the initially planned 1,103 maps were produced, and they have since been forgotten. Despite covering several urban fringe areas, such as Strasbourg and Reims, no mention of their use has been found. The problem they encountered was that classifying agronomic constraints, which might have seemed like a purely technical operation, implied taking a position on the land development model. (Arrouays *et al.*, 2022).

Soil quality assessment measures are still very sporadic

In the legal sphere, particularly in litigation cases that have been analysed, the concept of soil quality may be referred to as "agricultural interest", "agronomic interest", "agronomic potential", "agricultural quality", "agronomic quality", "land quality", "agronomic value" or "agricultural value". However, those involved in land use planning rarely explicitly define such quality. Instead, it is usually the "*Chambres d'Agriculture*" that provide local decision-makers with an agronomic diagnosis based on detailed field surveys and gelocated analyses (Boutet and Serrano, 2013), or even expert opinion, particularly through farmer

35. Annex III to Regulation (EU) No 1305/2013 of the European Parliament and of the Council of 17 December 2013 on support for rural development by the European Agricultural Fund for Rural Development (EAFRD) and repealing Council Regulation (EC) No 1698/2005.

surveys (Marseille *et al.*, 2019). These ad hoc soil quality assessment practices are then used in the preparation of urban planning documents in places such as Toulon, Nantes, Indre, Strasbourg and Poitiers. Various definitions are then proposed. Some are similar to soil science concepts and indicate that agronomic potential is the combination of climatic and geomorphological parameters, as well as the agronomic value of a soil–namely, "its structure and the presence or richness of plant nutrients found there, whether natural or cultivated" (Boutet and Serrano, 2013). Others, taking a heritage-based approach, consider soils with agronomic potential to be "soils that deserve to be passed on from the past in order to find value in terms of production potential for the future" (Balestrat *et al.*, 2011).

An example from Indre

In the Indre department, the Chamber of Agriculture has developed decision-making tools based on the soil map shown in Figure 3.2 and the associated 1/50,000 database (Antoni *et al.*, 2011). Maps of the agro-economic potential of soils, which are based on

Figure 3.2. Example of agronomic potential mapping in Indre-France. https://www.gissol.fr/donnees/cartes/un-extrait-de-la-carte-du-potentiel-agronomique-des-sols-de-lindre-2351 (accessed on 9/11/2024).

development constraints (e.g. stony ground excess water, soil texture and available water capacity), have facilitated the management of conflicts of use in peri-urban and rural areas during the expansion of activity zones or infrastructure development works within the framework of local urban planning (PLU). Mapping surface textures, water capacity, sensitivity to infiltration or crusting (the formation of a surface crust that promotes erosion) has also facilitated the establishment, delimitation and preservation of wetlands.

An example from Occitanie

A different approach taken in the former Languedoc-Roussillon region is well documented in the literature. In the early 2010s, an agronomic soil quality map was developed to provide local urban planning stakeholders in the five concerned departments with tools. The initial intention was to make these stakeholders, who are considered to have a primarily urban vision, aware of the need to preserve the best agricultural land for food production (Martin-Scholz *et al.*, 2013) by presenting this phenomenon in a quantitative and illustrated manner. This food-based approach defined soil quality based on its suitability for growing a variety of crops, particularly arable crops such as cereals and oilseeds (Clarimont *et al.*, 2021). The map was the result of collaboration between the Regional Directorate for Agriculture, Food and Forestry (DRAF), the Departmental Directorates for Territories (DDT), local authority representatives, INRA and CEMAGREF.

More generally, it has been established that soil quality is often overlooked when these zones are defined (Marseille *et al.*, 2019). Surveys report that "stakeholders admit their inability to consider the nature, depth and, above all, heterogeneity of urban soils" (Consalès *et al.*, 2022).

Regulating soil de-artificialisation and ecological restoration

Diversity of objectives and legal frameworks

When ecosystems reach a state of excessive degradation, they lose their resilience and capacity to return to, or promote an ecological trajectory towards their initial state. They may then transition to an alternative stable state. This abrupt conversion is often a response to minor, ongoing changes in one or more external factors that go beyond a threshold of degradation. This threshold is known as the ecological threshold (Spake *et al.*, 2022). The loss of biodiversity and ecological functions in this new stable state may justify implementing restoration practices to create a trajectory that leads the environment towards a reference state.

Ecosystem restoration is recognised internationally as a means of repairing damage, and it is also part of an approach to compensation aimed at achieving a world without net degradation. It addresses both biodiversity and land issues. This raises the question of how much consideration is given to degraded soil quality in this restoration objective.

French law encompasses a range of concepts with varying requirements. These range from the rehabilitation of an area to make it compatible with a particular use to the ecological restoration of environments within protected natural areas. One such concept is renaturation, which is currently under development and may involve a variety of technical processes. Regarding the objective of restoring functionality, only the de-artificialisation of soils, as set out in Article L. 101-2-1 of the Urban Planning Code, expressly refers to this. At the European level, the proposed directive on soil monitoring and resilience sets out the objectives of "regeneration" and "renaturation", which are aimed at restoring degraded soils to a healthy state.

Restoration ecology terminology

In the academic field of ecological restoration, a nomenclature developed by the Society for Ecological Restoration (SER) covers the diversity of objectives and corresponding trajectories, as illustrated in Figure 3.3. Restoration in the strict sense aims to reinstate all the characteristics of the reference ecosystem, including its species richness, composition, structure, and functions. Rehabilitation, on the other hand, focuses on restoring certain ecological functions or ecosystem services, or on the partial restoration of ecosystem attributes. A common example is the de-sealing of an area to compensate for sealing elsewhere. Once soil is sealed and has lost all, or in the best-case scenario most, of

Figure 3.3. Positioning of the various terms used in ecological restoration in relation to the expected objectives and trajectories (Jaunatre, 2012, based on Aronson *et al.*, 1993; Buisson, 2011).

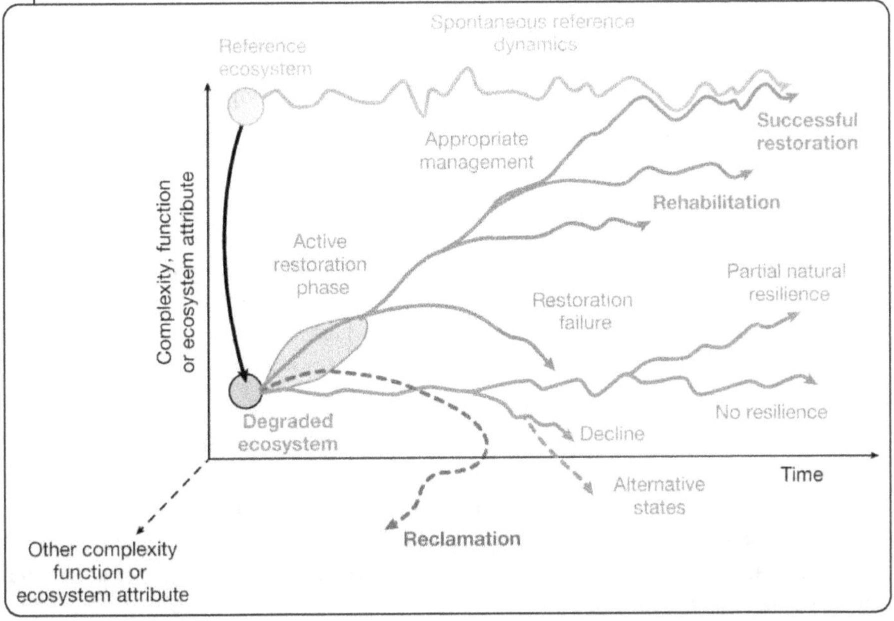

its functions, it is considered that these are effectively lost forever. The term "reclamation" refers to the stabilisation of soil, the protection of public health and safety, and the improvement of landscapes in the context of degraded industrial or mining land. Reclamation is also sometimes used as a synonym for reallocation (Aronson *et al.*, 1993). Finally, reallocation targets a different ecosystem to the one chosen as the reference for restoration and rehabilitation. This target is identified with the aim of improving biodiversity or the provision of ecosystem services.

Monitoring operations

The strategy for locating restoration operations is yet to be explored from both research and an operational perspective.

For example, the SCOTs (Territorial Coherence Schemes) provide for the identification of "preferential renaturation areas". The law nevertheless does not specify whether de-artificialisation will be considered once the work has been completed or when the ecological refunctionalisation of the soil can be observed. This question is crucial for preserving the environmental quality of the territory in question, given the considerable uncertainties surrounding restoration operations. In a study of French cases, Weissgerber *et al.* (2019) report on the difficulty of implementing compensatory solutions that lead to actual gains in biodiversity. They studied 577 ha that had been restored to compensate for artificial areas. Only 3% of these were artificial before the compensation work began, while 81% were semi-natural habitats, the ecological improvement of which has not been established.

It is difficult to predict how long it will take for the ecosystem to recover. For instance, the extremely slow resilience of Mediterranean pseudo-steppes necessitates monitoring over several decades. In the meantime, regularly assessing ecological attributes makes it possible to correct restoration operations if they take an undesirable trajectory, and to adapt the management practices that often accompany these operations.

The criteria for the success of a restoration project and how to measure it remain the subject of significant discussion and literature reviews, which may be generic or specific to certain environments (e.g. mines, forests or wetlands). In 2016, the Society for Ecological Restoration (SER) proposed an initial version of a generic framework for assessing restoration progress and success. Subsequently, this framework took the form of a "restoration wheel" (Gann *et al.*, 2019), which is composed of six categories of ecosystem attributes: absence of threats; external exchanges; ecosystem functions; structural diversity; species composition; and physical conditions.

Although most ecosystems perform multiple ecological functions, the vast majority of scientific studies evaluate the success of restoration based on only a few properties. The most commonly used properties are the soil's physicochemical characteristics (pH, organic matter, nitrogen, phosphorus, potassium and structural stability), the composition of the plant and edaphic communities (abundance and diversity of soil fauna and microbial communities), and sometimes the presence or absence of certain animal species or taxa (e.g. ants and birds).

Scientific studies should now incorporate all soil functions and analyse the role of different taxa in restoring multifunctionality. They should also extend their assessments of multifunctionality to various restoration operations in different ecological contexts. Long-term studies should evaluate the longevity of these positive effects, as well as the impact of the time lag between restoring ecosystem attributes and multifunctionality on the stability of restored ecosystems (i.e. their resilience, resistance and constancy) in the context of climate change.

Regarding engineered soils experiments, a considerable amount of literature has been identified. But at this stage, the conclusions are difficult to generalise due to the heterogeneity of the literature. Follow-up studies allow for case-by-case observations, but these are not comparable between studies and remain short-term given the timeframes involved in the processes. Furthermore, some operations may target only one function (e.g. water infiltration for de-sealing) with little information on concomitant changes to other functions. Nevertheless, creating green spaces after de-sealing is highlighted as a suitable way to rehabilitate soil functions. Table 3.1 shows examples of properties monitored following restoration of degraded soils and engineered soils. These indicators are generally associated with those used to monitor vegetation development.

Table 3.1. Properties studied for monitoring the restoration of degraded soils or the construction of constructed soils.

References	Objective	Physical Properties	Chemical properties	Biological properties
Ruiz *et al.*, 2020	Land reclamation (quarry)	Aggregates stability Bulk density	pH, P, K, SOC	Microbial biomass, enzymatic activities
GISQ Domínguez Haydar *et al.*, 2019	Reclamation (mine coal)	Texture, bulk density, morphology (origin of aggregates)	P, N, CEC, Ca ext., Mg ext., K ext., pH, OM	Communities of macroinvertebrates
Vincent *et al.*, 2018a and b	Constructed soil	Physicochemical characterization (not detailed)		Microbial communities (bacteria, fungi) Macroinvertebrates Micro-arthropods (including springtails) Herbaceous flora Activities (mycorrhizal colonisation, enzymatic activities, mineralisation capacity, macro-decomposers, meso-decomposers)

Ca: calcium; CEC: cation exchange capacity; SOC: soil organic carbon; ext.: extractable; GISQ: General Index of Soil Quality; K: potassium; OM: organic matter; N: nitrogen; P: phosphorus.

Territoriality of public intervention on soil quality

Complementarities between the European framework and local initiatives

The reviewed studies confirm the significant influence of international frameworks on the definition of environmental policies. In Spain, for instance, one article clearly demonstrates that the implementation of a soil erosion management policy is primarily driven by international (the 1993 Convention to Combat Desertification) and European (the Common Agricultural Policy [CAP] and the 2000 Water Framework Directive) legislation (Van Leeuwen *et al.*, 2019). Legally speaking, the implementation of soil governance covering a state's entire territory would initially be based on a top-down approach.

The results of the NORMASOL project (Billet, 2016b) are an important step forward in the debate on a legal approach to preserving soil quality.

Firstly, such a legal framework would remove the compartmentalised nature of the law, which is particularly evident in this area. Secondly, it would encourage the integration of soil quality standards into other areas of legislation, thereby promoting environmental consistency in decision-making processes (Hervé-Fournereau, 2009; Pieper *et al.*, 2023).

From an operational perspective, local authorities have high expectations regarding the accessibility of information. Several avenues have already been explored in this direction. Recent work (Consalès *et al.*, 2022) aims to develop project management tools based on a limited number of diagnostic indicators. These tools are intended to reduce costs and enable the use of proxies that do not require the intervention of soil scientists. The tools are also intended to work with incomplete data sets and produce semi-quantitative analyses using common data processing software. Other work highlights the types of actors to be mobilised at each stage of producing an urban planning document, including data producers, emphasising the roles assigned to each (Néel *et al.*, 2022). As part of the MUSE project, a methodological framework for internal use within Public Establishment for Inter-municipal Cooperation (EPCI) has been developed. This framework requires a cross-cutting view of land use, enables services to develop their skills, justifies urban development choices and provides tools for adapting regulations on buildable areas. It also enables the identification of priority areas for unsealing/renaturation as well as translating soil-related issues into land planning documents ("*orientations d'aménagement et de programmation*") (Néel *et al.*, 2022). Meanwhile, the DESTISOL project, based on documentary analysis and surveys, seeks to support decision-makers in optimising the balance between land use and occupation at the scale of a site where a development operation is being carried out (Blanchart *et al.*, 2018).

There is this room for manoeuvre in the regulations to take soil quality into account more effectively in certain elements of the PLU relating to the entire municipality, such as the environmental assessment, the presentation report and the Sustainable Development and Planning Project (PADD), as well as in more specific elements, such as the planning and programming guidelines, the regulations, the reserved locations and the classified wooded areas. Natural risk prevention plans (PPRN) and ZAPs could also incorporate soil quality considerations, particularly since protection under the latter procedure must be justified by general interests relating to production quality, geographical location, or agronomic quality.[36]

Territorial levels of governance

Work on soil quality governance emphasises the importance of territorialising public action at the local or regional level to strengthen soil protection. With this in mind, the proposed European Soil Monitoring and Resilience Directive provides for the definition of soil districts for governance purposes and soil units for monitoring purposes within each Member State. Examining the literature allows us to consider the possible options for establishing a relevant territorial network.

Existing entities, in particular watersheds

The main advantage of this option is that it prevents the creation of multiple decision-making areas. The watershed area is frequently referenced due to the close ecological links between water and soil dynamics, as well as many other issues such as agriculture, forestry, leisure, industry, urbanisation and energy. However, watersheds often cut across areas where the soil is homogeneous on both sides. From a soil science perspective, they are therefore not a relevant delimitation.

In Australia, soil governance has been established within regional Natural Resource Management (NRM) bodies. These bodies are responsible for the integrated management of all ecological resources. The boundaries of these NRMs are based on watersheds. In contrast, in the United States, the creation of soil conservation districts demonstrates that administrative rationality led to the formation of one district per county, despite initial intentions to consider the biophysical boundaries and spatiality of the soil phenomena to be managed. Additionally, the districts have the ability to plan, finance and implement projects, but were ultimately never granted regulatory powers, despite initial plans to do so.

Specific entity, pedologically homogeneous, or small agricultural regions (PRA)

Conversely, various proposals prioritise soil characteristics in the delimitation criteria, disregarding administrative boundaries. These proposals may involve soil mapping units (SMUs),[37] or a set of properties associated with climate, vegetation cover and land use, such as organic carbon, cation exchange capacity, pH, bulk density, texture and water

36. Art. L. 112-2c C. rur

37. SMUs are "portions of soil cover that share common characteristics in terms of landscape and soil distribution" (https://www.gissol.fr/donnees/carte-sur-le-geoportail-4789) and "in which the factors involved in soil formation are homogeneous (morphology, geology, climate and, in some cases, land use)" (https://artificialisation.developpement-durable.gouv.fr/bases-donnees/referentiel-regional-pedologique).

holding capacity. These properties could be combined with risk mapping (e.g. erosion and pollution), the management of which is linked to soils. French PRAs could form the basis of such a division, representing a compromise between an approach based on soil characteristics and a purely administrative one.

Urban areas

A novel approach to the functional perimeter of urban areas is proposed in an article by Donadieu *et al.* (2016). This approach includes not only built-up areas, but also "unbuilt areas (green and aquatic infrastructure) necessary for the resilience of inhabitants in the event of minor or major, temporary or chronic crises". The aim is to create an environmentally coherent and politically legible entity for planning purposes, based on the principle of protecting and enhancing undeveloped land and greening it where possible. However, no details are provided regarding how the remainder of the rural area would be divided.

Issue-driven governance

Several articles suggest avoiding the creation of a new administrative layer for soil governance and instead establishing a forum for strategic coordination between stakeholders on specific issues, such as erosion. Some propose defining a governance scale based on soil function (Schulte *et al.*, 2015). Others envisage creating "ecosystem services districts" (Heal *et al.*, 2001) to optimise service provision at the relevant scale and find compromises between the various possible uses of natural resources. These districts would be given coordination and information responsibilities, as well as planning responsibilities (i.e. arbitrating between potentially conflicting services) and fiscal powers. This option is analogous to the NRM bodies established in Australia. Finally, some articles emphasise the need to define a governance perimeter that corresponds to the relevant scales and value systems.

Compromise between harmonisation and appropriation

A balance must be struck between the need to integrate and harmonise soil quality approaches at different territorial levels and the flexibility necessary for stakeholders to take ownership of local soil preservation measures. The sustainability of an indicator depends on the balance achieved between stability and flexibility, which enables it to be used by different stakeholders in diverse spatial and ecological configurations and for different purposes. This debate also extends to the methods used to develop and measure indicators.[38]

Harmonisation initiatives

Numerous research programmes have aimed to harmonise indicator measurement methods (e.g. ENVASSO, ECOFINDERS and LANDMARK)[39] in order to ultimately compare

38. See Chapter 5, section "Measuring parameter and indicator values", p. 83.

39. ENVASSO: Environmental Assessment of Soil for Monitoring; ECOFINDERS: Ecological function and biodiversity indicators in European soils.

the results obtained by the various partners and ensure the stability of the methods used over time. However, given the diversity of available methods and the varying levels of complexity at which they can be applied, the chosen strategy takes into account the expectations and context of each study. The SIREN project report[40] highlights the wide variability of methods across EU Member States and paves the way for the short-term harmonisation of the first generation of soil quality indices (SQIs). This harmonisation is proposed while maintaining flexibility in the choice of methods and protocols within the framework of harmonised SQIs (i.e. limited standardisation). The possibility of differentiating assessment criteria according to regional context was also strongly advocated, reflecting the fact that soils, climates, and agricultural systems can vary considerably from one country to another.

The success of appropriating the indicators on which a conservation regime is based certainly depends on the scientific relevance of the indicator in describing a given situation, how easily it can be measured, whether the scale at which the indicator is applied is consistent with the manager's scale of action, the operating costs associated with the indicator and how comparable it is with other forms of measurement. However, success also hinges heavily on the social nature of its trajectory.

Social acceptance

When an indicator becomes established and is used in an institutionalised manner, enshrined in procedures and considered a benchmark for setting standards and assessing the effectiveness of public policies, it is said to be endowed with "new powers" (Bouleau and Deuffic, 2016). It helps shape the infrastructures and communities of practice around it, within which generations of participants become familiar with how it functions, its limitations, and the issues it addresses. This coordination work mobilises significant cognitive, economic, and political resources, leading to path dependency due to the prohibitive cost of questioning or adjusting the indicator regardless of its relevance (see Box 3.1). Bouleau and Deuffic (2016) also observe that the longer an indicator has been in use, the more cost-effective it becomes and the more expensive it is to change, due to the accumulation of long-term monitoring data.

Avoiding the rigidity of "problem closure", as defined by Forsyth (2007), poses challenges both scientifically and politically. This involves diversifying the scope of expertise that is considered legitimate for participation in scientific discussions. Rather than basing environmental governance on universal, predefined indicators, it allows for more flexible criteria that are interpreted on a case-by-case basis. These criteria take into account the local socio-environmental context and reflect on how they are defined. This may affect local actors politically, economically and socially.

40. SIREN: Stocktaking for agricultural soil quality and ecosystem services indicators and their reference values.

Box 3.1. RUSLE as a quantification of erosion.

The Revised Universal Soil Loss Equation (RUSLE) is one of the primary mathematical models used today for quantifying and predicting soil erosion. First published in 1992, it is the result of a series of corrections and adaptations to the Universal Soil Loss Equation (USLE). Following the Dust Bowl of the 1930s, the Soil Conservation Service of the United States Department of Agriculture (USDA) developed it to guide the soil conservation policy it implemented. The model considers precipitation patterns, soil type, topography, cropping systems, and management practices. A revised version of the model is still used in soil and water conservation programmes around the world today.

However, this equation has been widely criticised. Soil scientists such as Ernest Hallsworth (Hallsworth, 1987) and political ecology researchers (Blaikie, 2016; Forsyth, 2007) have argued that it is not "universal". The equation uses average rainfall figures which do not take into account the intense storms that cause most erosion in many countries worldwide. Furthermore, it does not account for soil formation or reformation rates, which are important parameters for assessing the acceptability of soil loss in certain regions. Despite attempts to adapt it to different local contexts, it fails to consider a significant proportion of local soil conservation practices.

Reports from recent FAO symposiums on soil erosion reveal a paradox: although the RUSLE equation is widely used by many stakeholders, including international development agencies and the European Commission, it is also widely criticised within these communities. Experts participating in these symposiums have criticised the equation for only considering diffuse water erosion and ignoring concentrated erosion, wind erosion, the effects of ploughing and harvesting-related erosion. These factors can be much more significant than diffuse erosion and are likely to become more serious in the context of climate change (FAO, 2019b). Nevertheless, due to the widespread use of RUSLE, little research has been conducted on these factors. The experts also note that RUSLE is often used on a scale much larger than that for which the tool was designed (FAO, 2019b). For all these reasons, RUSLE can produce uncertain findings and predictions that can have a significant impact on public policy, as they are taken seriously by decision-makers who are not familiar with the relevant scientific debates.

More reliable, comprehensive and better-adapted models have existed for decades (Fadil and El Wahidi, 2023). The continued use of RUSLE appears to be primarily due to the appeal of its simplicity and multiplicative nature. This is also due to the fact that many users are familiar with it, which facilitates data capitalisation and situation comparison. This helps stabilise understanding of the phenomenon and provides a useful summary of its causes and effects for managers. Once the "problem closure" stage has been reached (Hajer, 2005), whereby the problem has been "reduced" to a few seemingly simple and manageable parameters, the resulting definition tends to guide future research on the subject and select the types of expertise that will be consulted on the issue, thereby further reinforcing path dependency.

* Name given to a series of dust storms that occurred in North America in the 1930s. The Dust Bowl had a devastating impact on the ecology and agriculture of the affected regions, highlighting the importance of preserving soil quality in order to prevent and better withstand this type of hazard.

4. The dimensions of soil quality and health, and the choice of functions

Historical developments in terminology and concepts

In the biotechnical field, soil quality is a term that encompasses a wide range of notions and concepts which have been the subject of various studies. These include fertility, pollution, soil functioning, fulfilment of functions and multifunctionality, soil health, living soil, the provision of ecosystem services, the contribution of nature to people, to nature and to culture, resource sustainability, resilience, soil degradation and the threats thereof, soil value and even soil security. This proliferation of terminology reflects the difficulty of defining soil and its quality as objects of study. As an interface between different compartments and components, soil is strongly connected to other entities, making its delineation always debatable. This is why, rather than settling on a definition that would claim to put an end to the debate, the choice has been made here to shed light on the issues surrounding this terminological diversity and the tensions that shape it.

Internal and external factors

The use of different scientific terms to describe soil quality reflects various historical developments. In some cases, this usage may span several eras and undergo conceptual changes without any change in vocabulary. This applies particularly to terms such as "soil quality", "health" and "fertility", which are close to everyday language and have evolved in meaning over time according to research communities and users. In other cases, developments can be more clearly dated by the proposal of a concept, accompanied by a definition, which is adopted and used by a research community. Examples include ecosystem services in the late 1990s and soil security in the mid-2010s (McBratney *et al.*, 2014).

The evolution of scientific concepts relating to soil quality is linked to both external factors, corresponding to the place occupied by this notion in societal perceptions and concerns, as described in Chapter 2 (section "Diversity of perceptions of soils and their qualities", p. 25), and to internal factors, corresponding to scientific tools, whether technical or conceptual. Together, these two categories of factors explain the historical developments in the conceptualisation of soil quality.

Historical overview of the relative importance of systemic and analytical approaches

In ancient times, soil quality was understood in a holistic and metaphorical way, dating back to antiquity. Soil was considered as a whole entity, the various parts of which could only be understood in relation to the whole. It was viewed as a nourishing entity whose health was essential for the well-being of the human inhabitants of the Earth. For a long time, the study of soil was based on observing plant growth as an indicator of its health, most often in connection with agriculture. However, with scientific developments in the 18th and 19th centuries, soil quality could be analysed in a more atomistic way, based on its physical, chemical, and biological properties. Unlike the holistic approach, the aim was to develop an understanding of the soil as a whole based on the characteristics of its parts. A distinction was made between elements that were considered intrinsic to and representative of the soil type and elements that could be manipulated (often referred to as "manageable"), responding to management actions.

In the 20th century, soil science experienced significant growth, fuelled in particular by Russian pedology (Dokuchaev, 1883). Since then, pedology has explained the organisation of soils based on their profile descriptions, the structure of which results from the physical, chemical and biological processes of pedogenesis, using an integrated naturalistic approach. However, industrial developments in chemistry and mechanisation also made it possible to move from optimising soil quality through practices to substituting it with inputs. Soil was consequently reduced to a mere supporting role for plants and human activities. Imbalances caused by this industrial development led to an awareness of the risks associated with soil degradation. The main threat identified was soil loss caused by erosion linked to mechanisation, as exemplified by the devastation caused by the Dust Bowl in the United States in the 1930s. From the 1970s onwards, the accumulation of chemical residues, particularly in urban areas, has raised awareness of the dangers of soil pollution.

In line with the growing awareness of the importance of soil organic matter in combating erosion and its involvement in many other issues, scientific work is paying greater attention to the biological component. A systemic and multidisciplinary approach to soil is being adopted again, which is ultimately similar to the holistic approach of ancient times, but with a greater understanding of the interactions between the elements of the system. From the 1990s onwards, growing awareness of environmental limitations has led to the productive function of soil being considered less central to its study, with its other roles in relation to hydrology, climate, biodiversity, heritage and culture being highlighted instead. In the 2000s, the concepts of multifunctionality and ecosystem services, which had been developed earlier in other fields, emerged in the field of soil science. This systemic expansion continued into the social sciences and humanities, considering soil-human interactions on a global scale. These dimensions were integrated into the 2009 concept of soil security (Brauch and Spring, 2009), which was developed in relation to international issues of desertification. Improved knowledge of ecosystem dynamics and the importance of interactions between biotic and abiotic components, as well as the rapid development of biology, genomics and ecology tools, has led to a refocusing on the systemic dynamics of soil and a broadening of the dimensions considered. This has resulted in a shift in the concept of soil quality, with terminology related to ecosystem and soil health experiencing a resurgence in recent decades.

While these historical variations can be traced in broad global trends, significant differences emerge depending on the spheres in which the terms and concepts are adopted, sometimes in connection with social movements. The notion of soil health is more particularly embraced both by certain farmers (Richelle, 2019) and by international bodies, where it resonates with the "One Health" concept (which encompasses the health of humans, animals, and the environment, including soils). The notion of "living soil" has been associated with activist movements, particularly in agriculture and viticulture, as well as with groups of dissident scientists. For instance, during the 1950s, Afran (Association française pour une alimentation normale–French Association for Normal Food), aligned with the vitalist movement in medicine, criticised the extensive use of chemicals and its impact on food and health. The association published a periodical entitled "Sol-Aliment-Santé" (Soil-Food-Health) to promote techniques that foreshadowed organic farming. At the turn of the 21st century, terminology also varied between French- and English-speaking academic authors. French-speaking authors tended to develop the concept of soil quality in a more holistic sense, adopting the terminology relating to soil health only marginally. The latter is much more widespread in the English-speaking sphere.

A not-yet-stabilised distinction between soil quality and soil health

The historical dynamics outlined above demonstrate a correlation between the gradual expansion of the aspects considered in soil quality and the promotion of a more explicitly holistic concept of soil health, particularly within certain international institutions. Consequently, the meanings of these two terms remain significantly overlapping, and despite attempts at clarification since the early 2000s (Doran and Zeiss, 2000), they continue to be debated within the scientific community.

The first distinguishing features emphasise the greater importance given to soil biology and functionality in the concept of soil health (Karlen *et al.*, 1997). More recently, proposals have linked quality to inherent soil properties and health to performance in terms of ecosystem services, based on management methods implemented and the range of possibilities framed by inherent properties. This interpretation is similar to that proposed by Kibblewhite *et al.* (2008), whereby quality is considered as potential, and health is considered as the degree to which this potential is fulfilled, reflecting actual performance (which, by definition, is lower than potential and therefore limited by quality) and is partly linked to dynamic properties affected by soil management. According to this interpretation, illustrated in Figure 4.1, soil B is healthier than soil A.

More generally, the success of soil health terminology in scientific literature has been analysed as an intended metaphorical detour to raise public awareness, despite having no scientific basis. Some authors criticise the trend in academia of adopting terms that have been promoted for political reasons without them being sufficiently formalised and stabilised as concepts. They have proposed replacing the notion of soil quality with that of

Figure 4.1. Distinction between soil quality as potential and soil health as the degree to which this potential is realised.

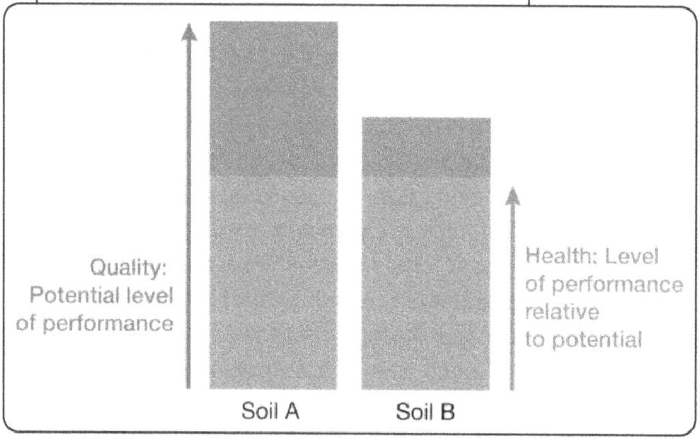

soil management quality, but this idea has never been further developed. Others emphasise the value of metaphor in science for renewing concepts. However, the difficulty of associating quantification approaches with such general concepts is an argument that recurs throughout the critical literature, whether we are talking about quality or health. Lehmann *et al.* (2020) point out that researchers have not yet resolved the "challenges of defining soil health in a way that allows for a universal quantitative assessment". They argue that soil health should be considered a fundamental principle around which knowledge should be developed in academia, rather than a quantity to be measured.

From soil quality to its ecological functions

Choosing an approach based on soil functions

Section 3.2.1 of Chapter 3 (p. 33, "Measures and economic values of soil quality"), showed that valuing soil quality too closely in terms of benefits directly perceived by users does not lead to its maintenance. In line with the changes to the definition of soil quality mentioned in the previous section, "Historical developments in terminology and concepts " (p. 59), the challenge today is to integrate the ecological functioning of soils more effectively into quality assessments. The decision to adopt a function-centred approach for this study was based on the interests and limitations identified below.

The concept of function is retained in the 2021 French Climate and Resilience Law.[41] This conceptual link enables the articulation between processes and services.[42] This approach

41. See Chapter 3, section "Legal criteria relating to types of use", p. 44.

42. See Chapter 1, section "Soil functioning and human activities", p. 8.

facilitates the valuation of ecosystem services while ensuring the preservation of all ecological functions. Given that an ecosystem service is underpinned by a variety of functions and that a function sometimes underpins a variety of ecosystem services, this approach covers the various dimensions of soil quality as comprehensively as possible.

Within the European Soil Observatory,[43] monitoring has so far been based on a degradation approach. Assessment is based on eight main threats to soil health: artificialisation, and, for non-artificialised soils, loss of biodiversity, loss of organic carbon, pollution, excess nutrients, compaction, salinisation and erosion. Soil is considered degraded when one of these threats reaches a critical level. This approach has met with resistance, partly due to the discouragement it can cause among stakeholders. This is because it is constructed in such a way that the risk of a negative diagnosis increases as knowledge about soil degradation progresses, which may discourage improvements in soil monitoring and tracking of soil functions. In this study, therefore, the decision was made to favour a soil health approach that is mobilising and based on ecological functions. Soil functions and degradation can be assessed using similar or even identical indicators.

Six ecological functions that reflect soil quality

The bibliography is incomplete with regard to the definition of the concept of "function" and there is considerable heterogeneity in the delineation of the different soil functions. Some sources consider functions and processes to be equivalent, while others do not. Nevertheless, there is a fairly broad consensus that a function is a bundle of interacting biophysical and chemical processes (Bünemann *et al.*, 2018) that enable the soil to function.

The methods used to identify the integrated actions that form a function, depend on the disciplinary perspectives and objectives of the various studies. Consequently, a multitude of lists of soil functions have been proposed in the literature. In order to understand how soil functions are addressed in the existing body of scientific literature, a textual analysis was carried out on 152 main articles studying these functions, the vast majority of which were published after 2013. These articles list from 1 to 32 functions, with a total of 658 different terms. Observing the most frequent terms in this corpus reveals two main types: action words (in descending order of occurrence: cycling, regulation, storage, production, sequestration, etc.) and object words (in descending order of occurrence: nutrients, water, soil, carbon, etc.).

On this basis, it was agreed that a soil function should be defined as the soil's action on an object or element. The names were grouped according to the most frequently occurring action-object associations found in the literature, with consideration given to the action of soil on itself (i.e. the formation and maintenance of its structure). This resulted in a list of six functions influencing the composition, structure, and flow of elements observable in soils (see Figure 4.2).

43. EUSO, https://esdac.jrc.ec.europa.eu/esdacviewer/euso-dashboard (accessed on 4/11/2024).

Figure 4.2. Scope of the six soil functions selected for this study.

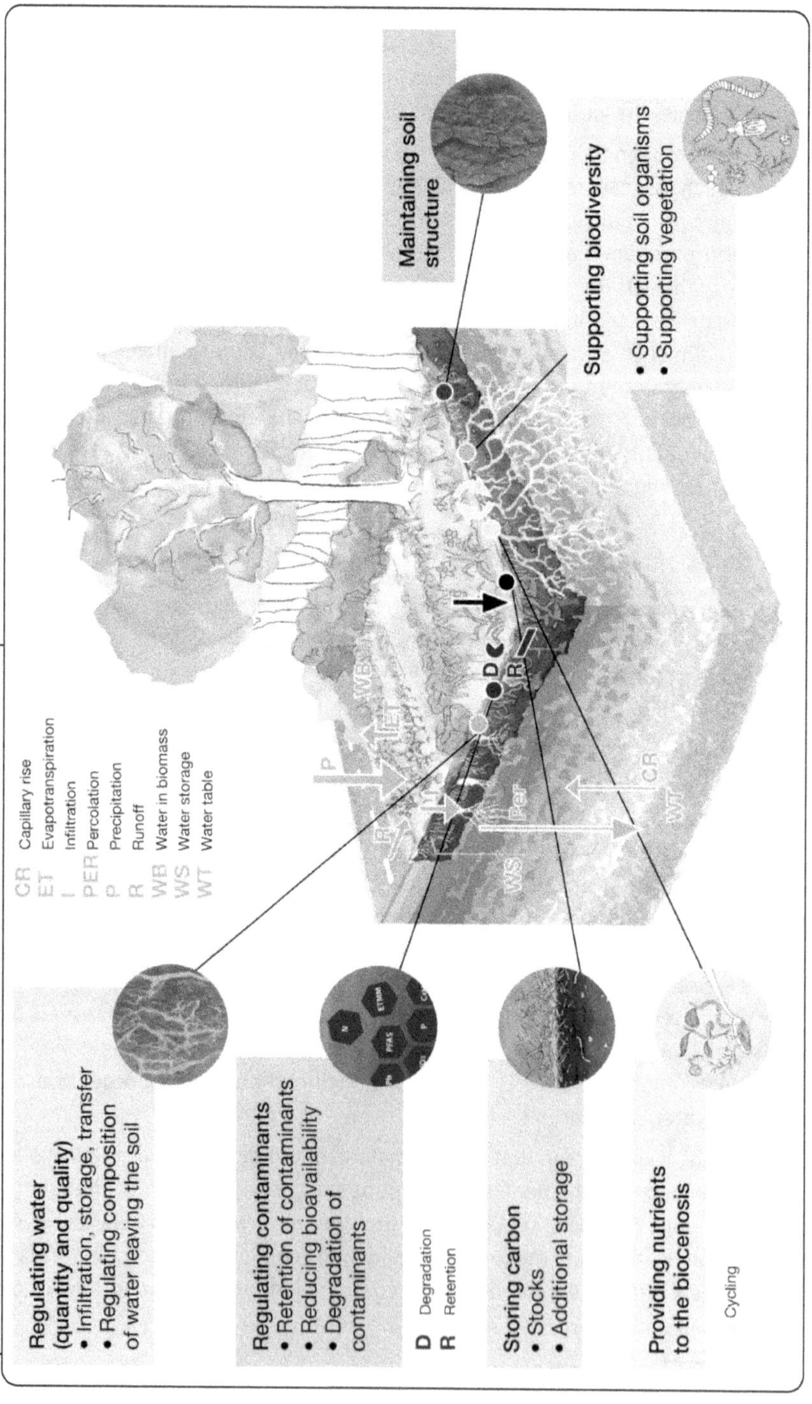

Table 4.1 presents the correspondences between this list of functions and the institutional frameworks that address soil functionality. On the one hand, there is the EU Soil Strategy for 2030 (COM[2021] 699), on which the proposed Soil Monitoring and Resilience Law was subsequently based. On the other hand, there are the provisions of the French Urban Planning Code resulting from the 2021 Climate and Resilience Law (Article L. 101-2-1).

Function "Supporting biodiversity"

Soil is one of the largest reservoirs of biodiversity on the planet. According to a recent study which took into account the estimated richness of all organism groups, the soil is habitat to around 60% of terrestrial biodiversity (Anthony *et al.*, 2023). To be considered a

Table 4.1. Tentative comparison between the selected functions in this study (see Figure 4.2), the outlined services in the EU Soil Strategy for 2030 and the listed functions in the French Climate and Resilience Law.

Services listed in the EU Soil Strategy	**Functions listed in this study**
Provide food and biomass production, including in agriculture and forestry	
Absorb, store and filter water and transform nutrients and substances, thus protecting groundwater bodies	
Provide the basis for life and biodiversity, including habitats, species and genes	
Act as a carbon reservoir	
To serve as a platform for human activities and to constitute an element of cultural heritage	Non-ecological function
Act as a source of raw materials	Non-ecological function
Constitute an archive of geological, geomorphological and archaeological heritage	Non-ecological function
Functions listed in the Climate and Resilience	**Functions listed in this study**
Agronomic potential	
Water functions	
Biological functions	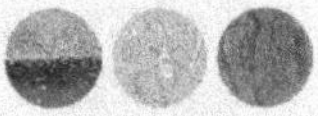
Climatic functions	

soil organism, an organism must complete at least one stage of its life cycle in the soil or one of its appendages, such as litter, decaying trunks, animal carcasses or excrement. Two sub-functions are distinguished because they can be associated with different indicators.

The sub-function "Supporting soil organisims" (i.e. non-plant organisms) is defined as the soil's ability to maintain or promote the diversity, abundance and efficiency of life forms (e.g. communities, populations, species, genes, molecules and enzymes) that interact with each other and generate biological and ecological processes.

The sub-function "Supporting vegetation" refers to the soil's ability to sustain the appearance of plants, their survival, the production of plant biomass, reproduction, resistance to stress, survival and natural or induced successional processes. This definition encompasses all land-based environments that are covered by plants, particularly those where productivity is not the primary expected ecosystem service.

Function "Storing carbon"

Carbon storage is the process by which solar energy and CO_2, captured by photosynthesis in plants, are incorporated into the soil. This storage provides energy for biocoenoses and helps to regulate CO_2 levels in the atmosphere (Wiesmeier *et al.*, 2019). There are two distinct sub-functions of carbon storage: one relating to the soil organic carbon (SOC) reservoir, representing the soil's current organic status, and the other relating to the additional storage capacity function, representing the amount of additional SOC a soil can store under certain constraints and conditions. It is also important not to confuse carbon storage with carbon sequestration. The term storage used here is generic and refers to all forms of carbon accumulation in the soil, whereas "sequestration" refers more specifically to the net removal of CO_2 from the atmosphere.

Function "Providing nutrients to the biocoenosis"

A nutrient is a substance necessary for living organisms that can be assimilated directly or indirectly without undergoing transformation. These include "major" nutrients, such as nitrogen, phosphorus and calcium, as well as elements needed in small quantities, or "trace elements", such as copper and zinc. Although nutrients are essential to biocoenoses, they cause pollution in excessive doses. Carbon is the structuring element of living organisms, but it is not considered a nutrient here since its autotrophic acquisition (through photosynthesis) does not involve the soil, and its heterotrophic fate (decomposition or stabilisation) is dealt with in the "carbon storage" function. The supply of nutrients depends not only on the quantity present, but on their form as well. Some forms are bioassimilable (Marschner, 2012), while others are not. It also depends on their availability over time and space.

Hydrological functions "Regulating water quantitatively and qualitatively"

As a volume extending from the Earth's surface to the parent material that has been little or not at all altered, the soil interacts directly with the hydrosphere. Its surface receives some of the precipitation and helps distribute these rainwater flows, supplying various reservoirs. These precipitations can then infiltrate, run off, evaporate or be stored in the soil, reaching the subsoil and possibly the water table through percolation. Water is omnipresent in soil in various forms that are more or less available to plants, such as being a component of soil minerals and living organisms, or circulating as free water under the effect of gravity in mesopores and macropores. Water can also be more or less immobile and accessible to plants due to the microporosity of the soil (Haslmayr *et al.*, 2016). Te Wierik *et al.* (2020) proposed a global quantification of flows based on annual land-based precipitation. 38% of precipitation returns to watercourses and the ocean through runoff; 2–12% reaches the water table through percolation; and the majority (50–60%) is temporarily stored in the soil before returning to the atmosphere through evaporation and evapotranspiration. However, this overall quantification masks considerable temporal and spatial heterogeneity depending on climate, land cover and use, and seasonality.

Quantitatively, soils regulate the flow of water reaching their surface, as well as that rising from the water table through capillary action. This function is related to the capacity of the soil to receive, store and transfer water.

Qualitatively, soils regulate the composition of water passing through them. This function relates to the retention and degradation of compounds carried by water that percolates into groundwater and/or aquatic environments. This qualitative function is closely linked to the quantitative function, as both are associated with the movement of water in the soil profile and on its surface. As it concerns the flow of materials associated with water, it also complements the function "providing nutrients to the biocoenosis", which relates to nutrient availability in the soil, and the function "regulating contaminants", which relates to the dynamics of contaminant transformation in the soil.

Function "Regulating contaminants"

Soils act as passageways for many substances and can play a regulatory role in this process. These include contaminants such as pesticides, trace metals, and pathogenic microorganisms, as well as excess nutrients, such as nitrate and phosphate.

The regulation of contaminants involves various processes, such as retention or degradation. This can be achieved through three complementary sub-functions, which are not necessarily effective in all soils or for all contaminants:

- the ability to retain contaminants (e.g. through sorption) helps preserve related ecosystems by limiting transfers but also contributes to the accumulation of these contaminants in the soil;
- the ability to reduce the bioavailability of contaminants (e.g. through sorption, complexation, or speciation change) (Lin *et al.*, 2016);
- the ability to degrade contaminants (particularly *through* the activity of soil microorganisms) (Wolejko *et al.*, 2020).

However, it should be remembered that although soils can regulate certain contaminants to some extent, when concentrations of elements are very high, this regulatory function can no longer be performed and the soil remains or becomes contaminated.

Function "Maintaining soil structure"

Soil structure is defined as the spatial organisation of voids (pores) and solids within soil. Maintaining this structure refers to the soil's ability to maintain or promote this organisation, allowing it to function chemically, biologically, and physically. This definition highlights the structure's dynamic nature. Thus, a distinction can be made between the state of the structure and its stability. The former refers to the state of porosity or aggregation at a given time, while the latter is an indicator of the structure's resistance to stress, such as the impact of raindrops, drought or compaction, whether natural or caused by human activities (e.g. machinery used in agriculture or forestry, or animals trampling the soil).

Relationships between functions

These six functions interact with each other in complex relationships that interlock at different scales. They are all mutually determining and determined.

For instance, the interaction between soil biodiversity and the abiotic conditions of its habitat plays a pivotal role in all functions. This can influence the supply of nutrients or be interpreted as a consequence of a certain level of nutrient bioavailability. Table 4.2 summarises the processes carried out by the main soil organisms and their influence on certain soil properties.

Maintaining soil structure is particularly linked to supporting biodiversity, including vegetation, and regulating water quantity. Structure is a fundamental element of organisms' habitat and the result of their activity. It determines the dynamics of water circulation in the soil and the nutrients and contaminants carried by the water. Conversely, structure evolves under the physical and chemical action of water.

Soil organic matter (SOM), which is composed of more than 50% carbon, influences most of the properties and functions of soil, such as water retention capacity, structural stability, nutrient storage and supply, and support for biodiversity.

As water circulates through the soil, it interacts with the different mineral and organic components of the soil to varying degrees, depending on how long it remains there. It participates in various processes, such as weathering, erosion, the mineralisation of organic matter, leaching, the transformation of nutrients, the adsorption of compounds, and so on. These interactions modify the physicochemical composition of the water circulating in the soil. More specifically, the soil solution and its interaction with organo-mineral components are fundamental to biomass growth.

Organisms cannot grow without nutrients. Therefore, providing nutrients is an important function for supporting biodiversity. But this function is also the result of biological activity, involving processes such as degradation, mineralisation and retention, to which organisms contribute. This function enables environmental resources to be transformed into a bioavailable form. It also depends on structure. For instance, a compacted soil horizon restricts plant roots from reaching deep into the soil, thereby reducing the reservoir of bioavailable water and nutrients.

Table 4.2. Effects of various soil organisms on main ecological processes, functions and properties (see function pictograms in Figure 4.2, p. 64).

Organisms	Main ecological processes	Functions	Example of soil properties
Bacteria, archaea	Nutrients cycle (N fixation, nitrification P solubilisation, etc.), soil carbon cycle, decomposition, aggregation, weathering, iron redox, degradation, regulation of populations		Cation exchange capacity, structure, structural stability, texture, redox potential...
Fungi	Nutrients cycle, decomposition, aggregation, contaminant degradation	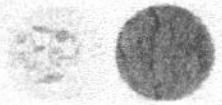	Cation exchange capacity, structure, structural stability, texture, redox potential...
Plant roots, lichens, bryophytes	Soil carbon cycle, weathering, infiltration		Structure, porosity, organic matter, pH...
Algae, diatoms, cyanobacteria	Soil carbon cycle, weathering	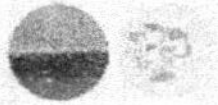	pH...
Protists	Regulation of populations		Microbial biomass, cation exchange capacity...
Nematodes	Regulation of populations, nutrient cycle	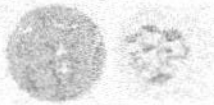	Microbial biomass and structure, net nitrogen mineralisation...
Arthropods	Decomposition, regulation of populations, material transport, infiltration, retention		Structure, (micro)structure, organic matter...
Earthworms	Decomposition, fragmentation, aggregation, material transport, infiltration, retention, degradation		Structure, porosity, clay organo-mineral complex, structural stability, organic matter...
Enchytraeids	Decomposition, aggregation		Structure, (micro)porosity, etc.

Soil with good water retention capacity increases the residence time of solutes within it, thereby enhancing the potential for contaminant retention and degradation. This capacity depends on the soil's physical properties, such as texture, depth, field capacity, water content and the presence of macropores, which often constitute preferential transfer pathways.

Approaches to assess soil multifunctionality

Concept of soil multifunctionality

The complexity of the interactions between these functions and their complementarities and antagonisms raises the question of how they can be taken into account in a comprehensive approach to soil functioning. How can we assess soil quality while considering its ability to transfer and retain water, and to store and recycle carbon?

The concept of multifunctionality emerged in the 1990s and gained prominence in soil science in the 2000s.[44] It encompasses a variety of issues and challenges. In everyday language, particularly in relation to land use issues, the term is generally used to emphasise the need to consider a broader range of factors, given the previously dominant focus on ecosystem services related to provisioning. The aim is to pay more attention to the regulating and cultural services provided by soils [see, for example, "*Entretiens avec les professionnels*" in Vincent and Blanchart, (2023)]. Manning *et al.* (2018) refer to this approach to the multifunctionality of ecosystems as "service multifunctionality", in order to differentiate it from "function multifunctionality". "Service multifunctionality" would have a more applied focus, consisting of evaluating the provision of services in response to human demand. Consequently, the weighting associated with each service may differ depending on the stakeholders involved.

The idea of function multifunctionality would aim to avoid any value judgements by objectively evaluating all the biological, geochemical and physical processes that regulate biogeochemical cycles. "Good" multifunctionality would therefore result from functions that give ecosystems efficient biogeochemical cycles characterised by rapid and effective nutrient recycling. However, this implicitly values soils with high rates of organic matter decomposition, mineralisation, primary production and water circulation (infiltration and percolation) while potentially neglecting functions such as water retention (e.g. wetlands), carbon storage (e.g. forests with little humus degradation and peat bogs) and biodiversity support typical of nutrient-poor environments (e.g. calcareous grasslands). This approach is by far the most widely represented in the analysed bibliography (86% of the 265 studies proposing a quantification of multifunctionality).

All functions are performed in all types of soil, albeit in different proportions and with different dynamics. Three main types of strategy have been identified in the literature to

44. See Chapter 4, section "Historical overview of the relative importance of systemic and analytical approaches", p. 59.

address the difficulty posed by the integration of functions. These strategies are described below, and their main advantages and limitations are summarised in Figure 4.3.

Assessing soil multifunctionality

Maximising multifunctionality

The first approach involves combining the assessments obtained for each function into a single multifunctionality measure, which we then seek to maximise. There are two main types of method that are most commonly used to aggregate functions into a single multifunctionality measure. The first method involves assessing the fulfilment of each ecological function in a binary manner (yes/no). Multifunctionality is then considered to correspond to the number of functions performed. This approach was used by Rabot *et al.* (2022) (Figure 4.3a), for example. The second approach involves quantifying the performance of each function using a score and then aggregating the functions by calculating the weighted average of these scores. However, this weighting is not only specific to the relevant stakeholders, but is also rarely implemented in practice. There is no reference framework that provides the relative weight of each function in an ecosystem, except when based on a target configuration chosen by the user as a reference.

Promoting diversity in functional profiles

One very common approach in the corpus (68% of the 265 studies proposing a quantification of multifunctionality) is to abandon the aggregation of functions into a single index. Instead, soil quality is evaluated by reading a functional profile composed of the degree to which the various functions are fulfilled individually (Figure 4.3b). This means that each soil unit is characterised by a set of functions, enabling the user to identify the unit's most suitable for certain uses or objectives (Baveye *et al.*, 2016). Moreover, this approach can identify synergies, compromises, and antagonisms between functions (Mouchet *et al.*, 2017).

To illustrate the spatial variability of soil functionality in this approach, Baveye *et al.* (2016) assigned a colour code to a typology of function clusters, which was then displayed on a map.

Preserving the rarest functional profiles

The approach to assessing multifunctionality based on observing high "rates" of function tends to favour soil types with the most dynamic functioning. To address this issue, a complementary approach can be adopted that takes into account the functional rarity of certain soils within a territory, i.e. the distinctiveness of the soil unit's functional profile. Due to the high level of uncertainty surrounding which functional characteristics will best suit future environmental conditions, favouring certain functional profiles over others reduces the potential for adaptation to these new conditions. This concept is known as the option value of biodiversity (Faith, 2021).

Figure 4.3. Diversity of possible approaches to soil multifunctionality.

	a) Number of functions	b) Functionnal profile	c) Functionnal rarity
Example of a graphical representation	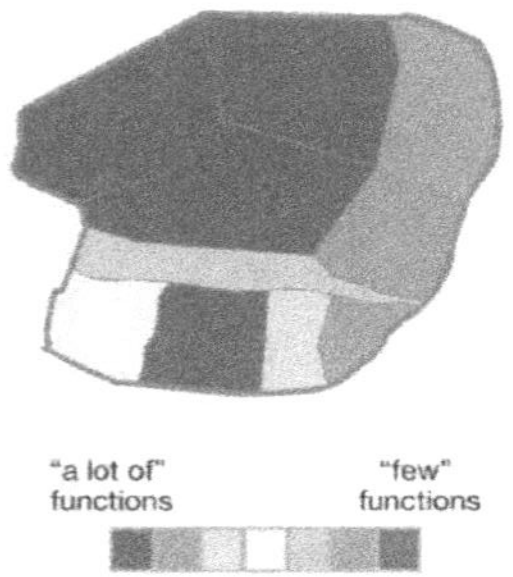 For an example of implementation, see Rabot *et al.* (2022)	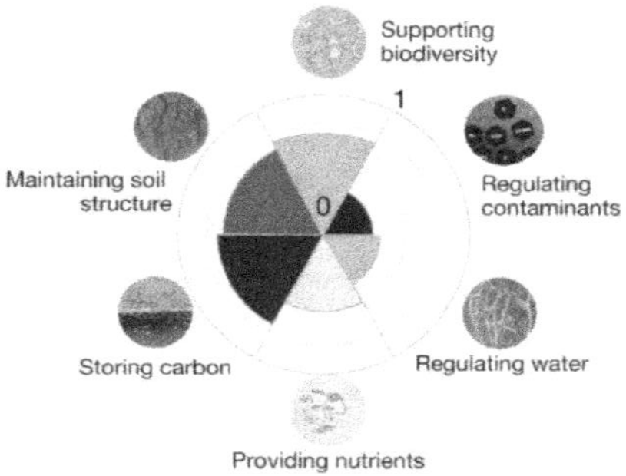 For an example of implementation, see Baveye *et al.* (2016)	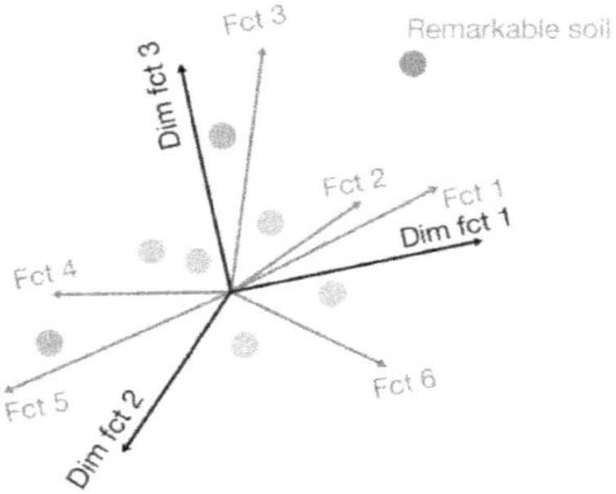 Fct = Function Dim fct = Functionnal dimension
Descrip-tion	Evaluation that aggregates different functions.	The evaluation of each function is based on the scoring of the values obtained for the indicators	Highlighting the rarest functional characteristics for a given territorial area.
Benefits	It allows summerising information based on a weighting of functions according to the user's interests. This aggregated information can then be visualised on a map.	Enables the functional characteristics of the soil at a given site to be identified, allowing choices to be weighed up in line with the objectives being pursued.	No preconceptions about the definition of soil quality. Corresponds to the ecological axiom that the rarest functionalities are key to ecosystem resilience.
Limits	The way functions are aggregated depends on prior choices. Aggregated information does not allow levers for improving management practices and uses to be identified.	To spatialise the information, a typology of profiles must be created.	Data-driven approach that incorporates biases associated with the constitution of data sets.

This proposal draws inspiration from the work carried out in ecology, particularly functional ecology, over the last three decades. Rather than assigning an absolute "functionality" value to species, the idea is to compare them with each other. This comparative approach enables us to identify species that are functionally irreplaceable within the ecosystem and, conversely, those that could be replaced by others if they were to disappear, a concept known as "functional redundancy" (Ehrlich and Walker, 1998). Methodologically, identifying functionally redundant species involves comparing their relative positions within an n-dimensional space constructed from the measured functions. This involves characterising the multidimensional distances between species and analysing them using methods such as principal component analysis (PCA) (Figure 4.3c).

Building on these findings, we propose a more exploratory approach to applying such a method to soils and their functions. This would enable soils that are more or less functionally similar to be identified in an n-dimensional space and the functional redundancy between soils to be characterised (i.e. the extent to which one soil shares the same functions as another). Notably, this would enable the value of soils with a particular and rare profile in terms of the functions they perform to be documented, even if they are not the highest-rated in terms of their multifunctionality. It would be important to preserve these soils precisely because of their rarity. Such information is useful for decision-making regarding protection. For instance, it is increasingly incorporated into the optimisation algorithms used to define biodiversity protection areas within a given territory (Pollock *et al.*, 2020; Pollock *et al.*, 2017).

This proposal to apply methods developed in ecology to highlight functional rarity in soils remains to be tested in practice, but it would introduce a neglected dimension to monitoring systems for soil quality and resilience.

Limitations of a function-based approach

In ecology, it is not justified to establish a hierarchy between functionally different soils, such as peatland soil (which stores significant carbon and water reserves but produces few nutrients), agricultural soil (which has rapid nutrient cycling but is sometimes less rich in biodiversity) and forest soil (which stores large amounts of carbon, provides few nutrients and has high biodiversity). However, any assessment approach involves making a judgement about a given soil situation. Thus, even if the concept of function is not primarily focused on supporting human objectives, many studies give it an artificial teleological dimension by assigning it a goal or purpose. Therefore, mobilising the concept of function in the context of assessment requires a choice to be made about the weight given to different functions.

Furthermore, the functional approach poses a risk to the preservation of the heritage of natural soils. Depending on their specific characteristics, natural soils perform different functions to varying degrees. A strictly functional approach could lead to the artificialisation of less functional natural soils, while favouring the creation of more functional soils. This could result in a loss of diversity in existing soils. In response to concerns about preserving soil diversity, the naturalness of soil remains a factor to be considered beyond functionality.

5. Assessment process

The diversity of perceptions and dimensions of soil quality, as well as the contexts in which it is assessed, makes it impossible to propose a universal reference framework for judging soil quality. The main aim here is therefore to highlight the main stages of the assessment process, emphasising the key considerations and methodological issues that determine the scientific validity of the results obtained, as well as the available resources to address these issues.

The proposed evaluation process comprises seven steps, some of which are optional. These steps are shown in Figure 5.1, with dots indicating the corresponding subsection number from those explained below. The approach begins with the objectives set out in public policy, which fall into three main categories, each of which leads to a different way of mobilising the elements of the assessment. Characterising quality involves carrying out a descriptive assessment of the soil without judging its quality. The preservation objective involves making a judgement on the situation being assessed, particularly with regard to threshold values. Finally, the objective of restoration is to define a target and corresponding target values. Based on these objectives, the figure shows the essential steps for assessing soil quality in solid lines and the steps that are activated on a case-by-case basis in dotted lines, depending on the objectives pursued and the levels and types of representations sought. It also highlights the steps at which users of the assessment are particularly responsible for making choices that they must explain, on the basis of which scientific experts can determine the parameters of the assessment.

Purpose of the assessment

It is important to make the parties involved clear about the assessment's purpose, as this will largely determine the subsequent steps. The implementation choices will not be the same, whether the aim is to describe soil characteristics or assess its potential in relation to a pre-established objective. This helps to explain the wide variety of reference frameworks proposed in the literature and some of the differences between them. Based on the diversity of situations encountered, this text proposes a classification of the main types of purpose.

Distinguishing characterising from assessing

Drawing up an overview

The most usual purpose of comprehensive soil quality monitoring is to inform public policy. This involves setting objectives and then examining their degree of achievement.

Figure 5.1. Steps for implementing the indicator system.

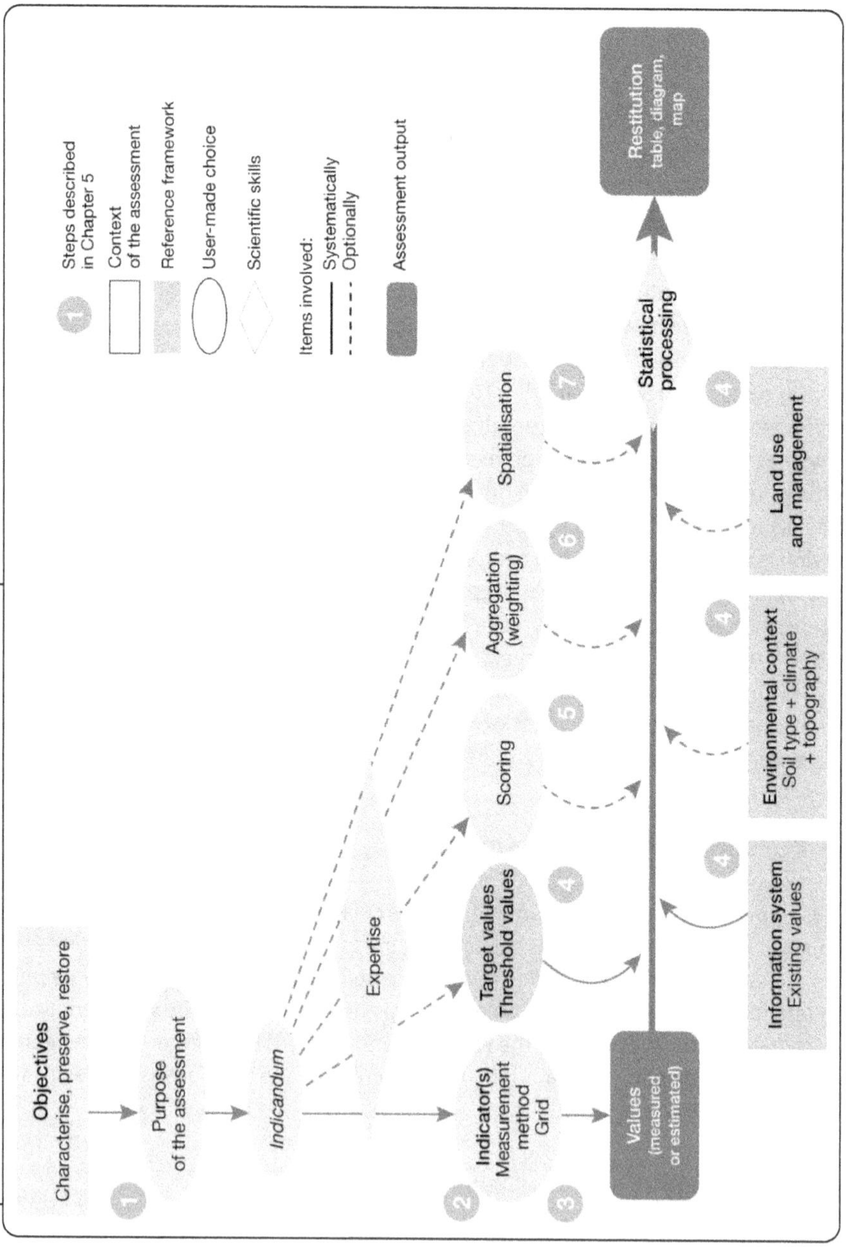

There are two ways in which this monitoring can be implemented, which differ in terms of their descriptive or normative nature. The latter involves making a judgement on quality.

The first approach involves characterising soil quality without making judgements. For instance, the global mapping of organic carbon stocks in soils carried out by the FAO[45] does not judge whether these stocks are low or high. Similarly, the 2011 map of forest soil pH in France (Antoni *et al.*, 2011) uses six pH classes ranging from very acidic to very basic soils without making any judgements.

The second approach involves combining an element of judgement with the information gathered. This is key to preserve soils or their functions against possible degradation, or to propose solutions to restore degraded areas. For instance, the European Soil Observatory[46] provides an online soil health dashboard which offers comprehensive monitoring of the percentage of the soil surface affected by at least one degradation process. It also provides a map showing impermeable soils and the number of degradation processes affecting non-impermeable soils. Transitioning from a descriptive characterisation to a normative assessment involves defining the criteria by which soil is considered degraded.

Supporting decision

Land managers must make decisions about land use and/or management practices. This involves assessing the suitability of the proposed uses and management methods in relation to the land's characteristics.

In land use planning, this means deciding how to distribute uses within a territory, ensuring the potential of the soil, as estimated by its characteristics, matches the desired uses, which are based on population needs. The assessment therefore covers all uses. The MUSE (Néel *et al.*, 2022) and SUPRA (Consalès *et al.*, 2022) projects are examples of this approach.[47]

When managing a site or plot of land for a specific purpose, the local diagnosis aims to evaluate the impact of the implemented management practices on the soil and optimise them based on the soil's potential determined by its characteristics. For agricultural use, for example, the LANDMARK project takes this approach using the Soil Navigator decision-making tool, as does the FOR-EVAL project[48] for forestry use. Scientific diagnosis enables the prioritisation of factors impacting soil quality and the recommendation of management measures.

45. https://www.fao.org/global-soil-partnership/pillars-action/4-information-and-data/global-soil-organic-carbon-gsoc-map/en (accessed on 9/11/2024).

46. https://esdac.jrc.ec.europa.eu/esdacviewer/euso-dashboard (accessed on 4/11/2024).

47. MUSE: Integrating soil multifunctionality into urban planning documents; SUPRA: Urban soils and development projects.

48. FOR-EVAL: Forest soil assessment.

Needs concerning mapping

It is also important to identify the potential spatialisation methods at the outset of the assessment, as these influence the statistical inference methods and sampling strategy.[49] For example, the results of an assessment can be described statistically in aggregate terms (mean, standard deviation, variance, etc.) at the geographical area level, or displayed on a map at the soil unit sampling level. These two options can be combined in various ways. They can also be cross-referenced with the decision of whether or not to include a value judgement, as discussed in the previous section. Table 5.1 provides examples to illustrate these possibilities.

Table 5.1. Examples of soil quality assessments by major types of purpose.

Finality	Statistics calculated for a geographical area	Spatialised statistics
Characterisation	Average organic carbon content of agricultural soils of France	Map of soil pH at a municipality scale
Judgment	Percentage of degraded soil in the EU	Map of the agronomic potential of soils at a region scale

Taking land use and management practices into account

Soil quality is the result of interactions between soil, climate and human activities over time. Different types of impact can be measured based on various soil properties and at different timescales and spatial scales. Some soil properties are more sensitive than others to human pressures. To highlight the effects of human interventions on soils and shed light on choices and decisions, research commonly distinguishes between properties that are sensitive in the short term and those that are sensitive in the long term. Thus, a distinction emerges in the literature between inherent, dynamic and/or "manageable" properties. The categorisation of these properties depends on the purpose of the study, particularly the timeframe of the decisions to be made and the way in which the type of use and management practices are considered. The distinction between these types of properties relies on the following elements:

- inherent properties, which do not vary over a time horizon of more than a few decades (e.g. roc fragments content);
- dynamic properties, which vary over a time horizon of less than a few decades, in relation to human activities or other changes in the ecosystem (e.g. bacteria/fungi ratio);
- manageable properties, which vary over the time horizon of interest to the users depending on their activities and objectives (e.g. nutrient content). This capacity to manage soil properties only refers to dynamic properties. Even though perennial properties are sometimes modified by human interventions (e.g. terracing, drainage and stone removal), they are not referred to as manageable properties in this context.

49. See below, section "Sampling issues", p. 83.

Therefore, the assessment methods will differ depending on whether the aim is to examine the impact of organic inputs on a plot over the course of a year; to evaluate the evolution of a plot's characteristics, considering all practices implemented during the leasing period; or to analyse the overall changes in agricultural soil conditions at the national scale, considering anthropogenic factors alongside climate change.

This point is important in terms of assessment because it involves considering the order in which information related to the pedo-climatic context(s), possible use(s), and type(s) of management practices is taken into account. Depending on the case, the same parameter may be considered either as a framing element or as the variable to be assessed. Figure 5.2 presents examples of fictitious situations chosen for their contrast in terms of assessment purpose. It illustrates how the elements are treated differently depending on the purpose of the assessment. This makes it possible to weigh up the soil functions enabled by different uses. Depending on the chosen purpose, it is also possible to consider a given use and vary only the management practices. The geographical frame of the study determines whether soil and climate are considered as constant parameters (in a very local study, depending on the spatial variability of soil type) or variables (at the level of a country or continent, with different types of climate).

It should be noted that the degree of detail considered for each of these categories (soil and climate conditions, uses and management practices) may be highly variable depending on the objectives of the study. Soil and climate conditions are described in numerous classification systems of varying degrees of detail. Examples include the Köppen climate classification (Beck *et al.*, 2018) and for soils, the WRB[50] or the more widely used French Référentiel Pédologique (Baize and Girard, 2009). Regarding land use, the large-scale land cover classification (OCS GE)[51] provides a nested nomenclature for France, offering a choice of 4 or 17 land use categories (e.g. agriculture, roads, and abandoned areas) and 2, 4, or 14 land cover categories (e.g. built-up areas, water surfaces, and conifers). In the agricultural sector, for example, the available typologies refer to practices (e.g. ploughing or no-till farming) or production systems (e.g. conservation agriculture), some of which are backed by certifications (e.g. organic farming).

Choice of the *indicandum* and of the relevant indicators

An indicator is used to obtain and transmit information about an object or phenomenon of interest, known as the "*indicandum*". The relationship between an indicator and its *indicandum* can take various forms. For instance, it may comprise representational links connecting the whole to some of its parts, or causal or correlative relationships that can be expressed mathematically. Selecting a subset of parameters always involves a reductionist approach that favours certain dimensions of the *indicandum* at the expense

50. https://www.isric.org/explore/wrb (accessed on 31/10/2024).

51. https://artificialisation.developpement-durable.gouv.fr/bases-donnees/ocs-ge (accessed on 31/10/2024).

Figure 5.2. Examples illustrating the diversity of methods for incorporating the pedoclimatic context, land use and management methods into soil quality assessments, based on three fictitious situations.

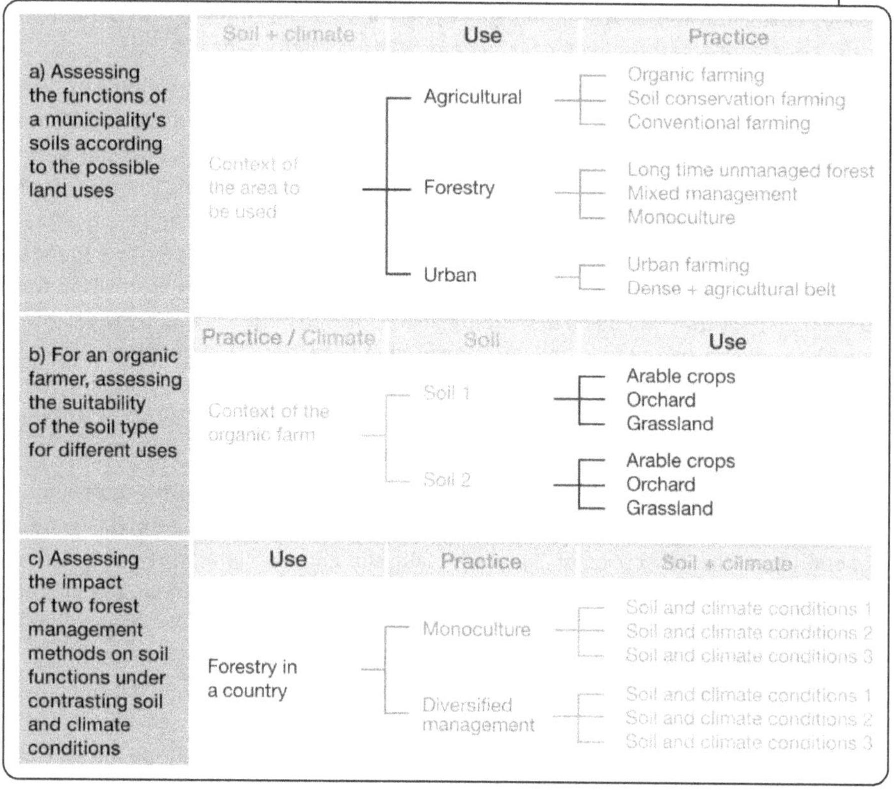

of the whole. Social science research shows that these initial choices must be transparent and documented, and ideally shared by all parties involved in constructing and using the indicator system (Czúcz *et al.*, 2021).

Defining the indicandum

A review of the literature on soil quality reveals that indicators are generally chosen without any clear rationale for the *indicandum* they are intended to represent. Chapter 3 ("Compromise between Harmonisation and Appropriation", p. 56) highlights a strong path dependency in this regard, resulting from working habits, the accessibility of tools and data, and the popularity of the most commonly used indicators. However, clearly defining the *indicandum* is an essential step in rationalising the selection of indicators.

Studies addressing soil quality in relation to public policy draw on a variety of conceptual frameworks, identifying different types of indicators (or objects of interest). One such framework is the Drivers-Pressures-State-Impacts-Responses (DPSIR) model (Niemeijer and de Groot, 2008), which categorises determinants (here understood as very generic factors of change, such as population growth), pressures, states, impacts and responses in ecosystem dynamics. In public policy and economics, threats (pressures) or the level of degradation (impacts) are often used as indicators, with erosion being a particularly prominent example. For instance, the soil health barometer developed by the European Soil Observatory is based on eight sources of degradation: erosion, compaction, salinisation, loss of organic carbon, loss of biodiversity, pollution, excess nutrients and artificialisation.[52] Other studies use the ecosystem services cascade conceptual framework[53] to categorise the condition, functions, services and benefits derived from ecosystems.

In connection with the 2021 French Climate and Resilience Law, this study decided to consider soil ecological functions as *indicandum*, which enable the most comprehensive link between soil properties and ecosystem services.[54]

Selecting of a coherent and effective set of indicators

The next step is to select the indicators that best represent the *indicandum*. This selection process is based on the assumption that variations in these indicators reflect variations in the *indicandum* itself. Scientific literature sets out various recommendations for identifying a minimum set of indicators (minimum data set, MDS) that capture this variability. The effort to minimise the number of indicators responds to practical concerns about the cost of the study and how easy the results are to read and interpret, as well as scientific concerns about exhaustiveness (i.e. the integration of the physical, chemical, and biological processes related to the *indicandum* being assessed) and non-collinearity or redundancy, to avoid unbalancing the relative importance of the measured dimensions. In the 1990s, the first work on formalising soil quality indices drew on the MDS concept introduced by Larson and Pierce (1991) to designate the minimum set of basic indicators to be considered when assessing soil quality. However, there is no consensus on which indicators should be included in the MDS, and the very process of identifying a valid MDS in all contexts is questioned in light of the diversity of approaches presented above under "Purpose of the assessment" (p. 75).

When the objective is to compare different options (e.g. land use or agricultural practices), the selected indicators must be sensitive to these options. In this case, the set of indicators is not chosen to cover all aspects of soil quality; rather, it is chosen to distinguish between the options under consideration. For instance, if the assessment aims to evaluate tillage, the analysed indicators will focus on characteristics most likely affected by tillage, such as compaction, structure, hydrological functions and the abundance and diversity of soil organisms. Contaminant content is less directly related to tillage and therefore not considered. In other words, even if an indicator provides information on an

52. https://esdac.jrc.ec.europa.eu/esdacviewer/euso-dashboard (accessed on 4/11/2024).

53. See Chapter 1, section "Soil functioning and human activities", p. 8.

54. See Chapter 4, section "Choosing an approach based on soil functions", p. 62.

essential element of soil quality, it may not be selected for the specific study in question if its value does not vary according to the options under consideration. Similarly, an indicator that discriminates between the available options but is unrelated to the aspects of soil quality to be assessed would also not be included among the selected indicators.

The selection of indicators must also consider the performance level of each potential indicator and its associated measurement methods. The performance conditions of these indicators are studied in scientific literature dealing with indicator-based approaches, preliminary work on indicator systems (particularly bioindication), methodological guides developed by national or international bodies involved in environmental reporting, and documents presenting feedback on experiences related to the implementation of indicator systems. This work brings together the criteria conducive to the operationalisation of scientific information—that is, its appropriation and use in public or private decision-making processes. These sources provide benchmarks and recommendations for this indicator selection. These recommendations can be grouped into four main categories according to the credibility-salience-legitimacy (CSL) approach, to which the criterion of feasibility is sometimes added (CSLF):

- **Credibility** refers to the scientific relevance of the indicators used to represent a phenomenon or object of interest. It is primarily assessed on the basis of the relationship between the indicator and the *indicandum*, as well as the characteristics of its measurement. This relationship must therefore be the subject of scientific consensus and as direct as possible. The measurement of the indicator must be explicitly specified to ensure accuracy and reproducibility;
- **Salience** refers to the relevance of the information provided by the indicator for public and private decision-making. For example, the indicator should be sensitive to management practices or changes in land use that are to be assessed. Above all, the indicator(s) in the indicator system must be consistent with the issues and needs of users. The information provided must be understandable and interpretable in order to facilitate diagnosis and decision-making. To achieve this, it must be possible to account for variations in environmental conditions, as well as user actions;
- **Legitimacy** of the process of selecting and constructing indicators for users of the indicator system reflects acceptance of the indicator, its measurement and interpretation, by various stakeholders. This is based on consideration of the diversity of users and their legitimate interests in the process of defining, selecting, and interpreting indicators;
- **Feasibility** refers to practical conditions for using an indicator, such as the suitability of the measurement method for available material, financial or human resources, or the availability of suitable spatial and temporal resolutions in databases for regular monitoring.

Concept- or data-driven selection

The approaches proposed in the literature for selecting a relevant set of indicators fall into two main categories: conceptual model-based approaches and data-driven approaches based on the statistical characteristics of the dataset.

Concept-based approaches have been formalised into methodological frameworks such as the Comprehensive Assessment of Soil Health (CASH; Fine *et al.*, 2017) and the Soil Management Assessment Framework (SMAF; Andrews *et al.*, 2004).

The CASH methodological framework offers three sets of indicators (packages), depending on the cost/accuracy ratio agreed by the user: basic, standard and extended. The choice is also guided by recommended applications corresponding to the context: "basic" for field crops, "standard" for organic farming and "extended" for urban soils. Additional analyses are recommended in specific situations, such as heavy metal contamination in urban soils, private gardens, playgrounds, and brownfield sites.

Data-driven approaches use descriptive statistical methods to identify the most explanatory indicators of variability in multiple datasets. These are multidimensional exploratory data analyses, the method of which depends on the type of data describing the situation (quantitative, qualitative, or both). Examples of these multivariate analyses include Principal Component Analysis (PCA), Redundancy Analysis (RDA) and Factor Analysis (FA). There is still disagreement among authors on a technical level about how to judge the relative importance of different indicators in decision-making. The differences concern which mathematical elements should be considered, as these are produced by calculations. These methods have the advantage of neutrality of choice, subject to selection biases associated with the characteristics of the initial dataset. They are sometimes described as *"a posteriori"* (Griffiths *et al.*, 2016) because the indicators are selected from a set of candidate indicators that have already been measured at the study sites in order to constitute the initial dataset.

However, it should be noted that these scientific data selection strategies are often supplemented by criteria relating to the existence or availability of data. The above approaches are sometimes unnecessary, particularly when the available dataset is limited.

Measuring parameter and indicator values

Once the indicators required to qualify the *indicandum* have been selected, obtaining their values can present significant methodological challenges. Indicators may be more or less directly linked to a measurable quantity. The methods used to obtain these values are also very diverse. They may be visual or instrument-based, involve a model or rely on a signal captured by proxy-detection or remote sensing.

Sampling issues

Soils, and therefore their morphological, physical, chemical and biological characteristics, vary greatly over space and time. There is no ideal solution for accounting for this variability, and any measures implemented will necessarily involve compromises, for example between accuracy and cost, or between responding to specific needs and ensuring comparability. Technological developments in recent decades, particularly in remote sensing,[55] Geographic Information Systems (GIS) and modelling, have led to more reproducible and rapid methods of soil quality mapping. These technologies allow data to be collected, analysed and visualised at different spatial and temporal scales.

Preliminary clarifications

In line with the purpose mentioned above in "Purpose of the assessment" (p. 75), implementing a soil quality assessment requires the clear explanation of the following elements:
- **Target universe** or universe of interest: this describes the population or area to be represented and sampled, with spatial and/or temporal limits, and possibly exclusions specified. For example, this could be the topsoil of all cultivated land in a region, or soil up to 1 m deep in forest areas of a country;
- **Area(s) of interest**: the part(s) of the target universe for which separate results are to be reported. These may be subdivisions of the target universe, such as land use types, administrative boundaries or possible soil districts;[56]
- **Target variable(s):** the variable(s) to be determined for each sample unit. For example, soil properties or soil quality indicators derived from measured properties;
- **Target quantity:** a combination of a domain, a target variable and a target parameter. For example, in the case of organic carbon, this would be the average content (parameter) in the 0–30 cm soil layer (target variable) of forest soils in France (domain);
- **Type of result:** either qualitative or quantitative. With a qualitative result, the inference method is testing, classification or detection. An example of this would be detecting the presence of carbon-rich soils in agricultural soils in France. With a quantitative result, the inference method is estimation or prediction.

Support definition

Depending on the objective(s), sampling is carried out in space and in the vertical dimension (for the latter, it is either based on heterogeneous stratification according to soil horizons of varying depth or on homogeneous stratification according to soil layers of constant depth, regardless of the actual link between a layer and a soil horizon). The set of measurements that constitute a sampling unit are referred to as a "support".

For example, the LUCAS Soil sample support is a 4 × 4 m block, in which a composite sample consists of five subsamples taken from a depth of 20 cm (Orgiazzi *et al.*, 2018). The sample thus represents the average of the block. Within the Soil Quality Monitoring Network (RMQS) framework, a composite sample comprises 25 individual samples taken at depths of 0–30 cm and 30–50 cm and spread over a 20 × 20 m block.

Various methods have been formalised for defining the support, the relevance of which depends on the expected result. De Gruijter *et al.* (2006) provide a detailed overview of the key considerations. As with the selection of indicators described above in "Selecting

55. See below, section "Advantages and limitations of some measurement methods", p. 88.
56. See Chapter 3, section "Territorial levels of governance", p. 55.

of a coherent and effective set of indicators", there is a dichotomy between model-based approaches, which simplify the variability of the phenomenon under study by formulating expert hypotheses, and design-based approaches, which are calibrated using the variability of the available data. For estimating global quantities such as averages and totals (e.g. the percentage of degraded soils in an area), it is accepted that probabilistic sampling approaches with design-based inference are more effective than model-based methods, providing valid and unbiased estimates of the associated uncertainties. The most appropriate sampling plans for spatial predictions (mapping) are based on various estimation methods known as kriging (Wadoux *et al.*, 2019). These methods involve optimising the variation of criteria related to model calibration. In sampling plans for monitoring, however, spatial variation is not the only factor to be considered, as temporal variation must also be taken into account. This makes the approaches to be considered more complex.

Implementation

To implement a sampling plan that aligns with the recommendations established in the literature while keeping data collection and management costs affordable, it is sensible to utilise existing soil data and ensure it is accessible and usable. Building on this, the sampling objective could be to describe situations that are not or poorly represented in the initial data more accurately, or to increase the sampling resolution, while ensuring the diversity of sources is articulated in a meaningful way. In France, for example, the data collected as part of the GIS Sol programmes encompasses various approaches.[57] The Soil Inventory, Management and Conservation (IGCS) programme primarily aims to produce maps rather than provide overall statistics. Consequently, sampling has not been designed to meet the constraints of probabilistic sampling theory, raising questions about quantifying the associated uncertainties. Only the observations collected as part of the RMQS meet these criteria, but the resolution remains too low to provide information at local scales. Therefore, efforts are still needed to build ad hoc datasets for a robust statistical assessment of soil quality at fine scales.

Quantifying uncertainties

Lastly, projecting the sample's measurements onto the reality we are seeking to observe carries uncertainties that can be quantified. Assessing these uncertainties is crucial for informed decision-making. For instance, regulations govern certain land uses in relation to pollutant content. Taking the uncertainty of the estimated average into account when verifying compliance with the standard has important consequences for the user. When the average is just below the standard, there is a high probability that a significant proportion of the population will exceed it. Therefore, it is essential to associate a margin of error with an estimate.

57. See Chapter 6, Box 6.2, p. 121.

Relationships between indices, indicators and measured quantities

The vocabulary relating to properties, parameters, indicators, indices and other descriptors is used inconsistently throughout the literature. For our study, we have adopted the convention illustrated in Figure 5.3 of reserving the term "indicator" for a meaningful attribute in relation to the *indicandum* that we are seeking to characterise or evaluate. Thus, a measured or estimated elementary quantity can also be an indicator (as with indicator i_2 in Figure 5.3), provided it is indicative of something. It is this representational capacity that gives the measured quantity its nature as an indicator. The measured or estimated elementary quantity is most often called a "property" when linked to a dynamic process or a "parameter" when used in a calculation or model. In Figure 5.3, it is named P.

In some cases, the indicator is calculated from several measured variables (see indicator i_1 in Figure 5.3). For instance, maximum available water capacity is often used as a physical indicator of soil hydrological functionality, and thus of soil quality. This value can be estimated using parameters such as soil texture, structure, and organic carbon content.

When several indicators representing different dimensions of soil quality are used, their values can be aggregated[58] into a single index. However, the term "index" is also often used in literature to refer to a composite measure of a single dimension of soil quality. In biology, for example, numerous indices have been developed to reflect a state (e.g. taxonomic richness) or a structure (e.g. the Shannon diversity index). Similarly, in chemistry, pollution indices are developed to account for overall pollution levels by aggregating substances. Mazzon *et al.* (2021) use the term "simple index" to differentiate them from complex indices, such as the SQI (Soil Quality Index) or the SFI (Soil Function Index). Box 5.1 provides some examples of simple indices, while the section "Multi-criteria aggregation forming a soil quality index" (p. 101) later in this chapter discusses complex indices obtained through an aggregation process. Finally, an even broader aggregation can be envisaged, integrating vegetation and climate as in the SI (Suitability Index) proposed by AbdelRahman (2023) or ecosystem quality indices, of which soil is a component. These indices are, however, beyond the scope of this study.

Some measurement methods quantify complex information directly. This is particularly the case for signals captured by proxy-detection or remote sensing[59] and translated into indices by modelling. These indices can reflect some aspects of soil quality; for example, the state of vegetation can be reflected by the Normalised Difference Vegetation Index (NDVI). While this complex information can be considered in itself, it can also be interpreted using algorithmic deconvolution to estimate the values of basic parameters or indicators. This contributes to the estimation of indicators and indices constructed by aggregation.

---

58. See below, section "Multi-criteria aggregation forming a soil quality index", p. 101.

59. See below, section "Advantages and limitations of some measurement methods", p. 88.

Figure 5.3. Terminology used in the study on the distinctions between index, indicator, and property or parameter.

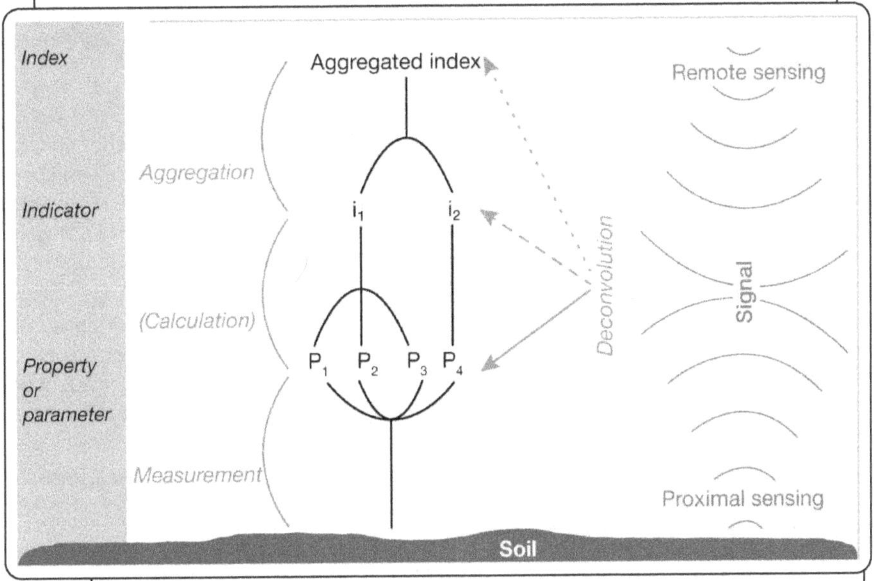

Solid arrow: most common use of the signal in proximal or remote sensing; Dotted arrow: less frequent use.

Box 5.1. Examples of indicators expressed as simple indices that incorporate only one dimension of soil quality.

Quantification of biodiversity structure

The Shannon diversity index expresses the diversity of a specific community. It takes into account both the number of species (specific richness) and the distribution of individuals within these species. It reflects the community's heterogeneity by taking into account both the number of observed species and their relative abundance. A higher index indicates greater diversity, but it is strongly influenced by the presence of rare species (low relative abundance).

The Simpson index quantifies diversity by calculating the probability that two randomly selected individuals belong to the same species. Consequently, it gives more weight to abundant species than to rare species. It is lower when the community is more diverse.

Quantification of biological activity

Litter on the soil surface is an important resource for decomposing organisms.

The litter index classifies this litter according to its state of degradation into two categories: fragmented litter and skeletonised litter (of which only the veins remain). The ratio of the dry biomass of skeletonised or fragmented leaves to the total litter biomass is then used as an indicator of decomposers' activity.

Box 5.1. (following)

Numerous microbial indices are also reviewed by Bhaduri *et al.* (2022), such as the *Biological Index of Fertility* (BIF), *Microbial Index of Soil* (Mi), *Enzyme Activity Number* (EAN) and *Soil Biological Fertility Index*.

Quantification of the degree of soil disturbance

The structure index of nematode communities depends on the relative abundance of several families of free-living nematodes (bacterivores, fungivores, omnivores and predators). It is calculated using a weighting system that considers the relative importance of these families compared to what would be expected based on a reference structure trajectory. It reflects environmental stability: the higher the index, the less disturbed the environment and the more complex the soil food web.

The Maturity Index (MI) and Plant-Parasitic nematode Index (PPI), which were proposed by Bongers (1990), are based on the abundance of nematode populations that are differentiated by their functional characteristics. MI values are inversely proportional to the extent of soil disturbance, whereas PPI values are positively correlated with soil disturbance. Changes in the relative abundance of nematode feeding on bacteria or fungi reflect changes in decomposition pathways. This is assessed by measuring the Decomposition Pathway Index (DPI): the higher the DPI, the more decomposition is controlled by bacteria.

Quantification of the degree of anthropisation or artificialisation

The Natural pollutant attenuation capacity of urban soils (NAC) is based on the following parameters, which are likely to control the fate of pollutants in soils: % Corg, % clay, bulk density, pH and total N. The nature of land use impacts it (in ascending order: parks < schools < woods < residential areas < traffic networks), and Wang *et al.* (2015) have shown that it correlates with the age of artificialisation. These authors propose it as a means of conceptualising soil recovery capacity and monitoring the pollution mitigation capacity of urban soils.

Quantification of pollution

A specific feature of Trace Metals and Metalloids (TMMs) is that there are many of them, yet they are considered a single parameter due to their similar or cumulative effects. To this end, specific indicators have been developed to estimate the degree of soil contamination. The pollution load index is defined based on the concentration of each metal measured and a reference content for that element. Calculating these specific indicators requires each element to be analysed in relation to geochemical backgrounds.

Advantages and limitations of some measurement methods

Different measurement methods can produce different values for the same indicator, so it is essential to specify the method used to avoid misinterpreting the results. For indicator values to be comparable between situations or territories, calculation and measurement methods (including sampling and sample processing protocols) must be harmonised. Some of these methods have therefore been standardised through Afnor standards in France and ISO standards at the international level. Standards may also be developed by learned societies, international organisations (e.g. the Global Soil Laboratory Network–GLOSOLAN, FAO), or even non-governmental organisations (e.g. the World Biodiversity Association).

There are also multiple methods for measuring basic parameters. For example, two standardised methods exist for measuring total organic carbon: dry combustion using an elemental analyser (ISO 10694 [1995]; SA08), which is favoured by ICP Forests, WRB and FAO (FAO, 2019a); and wet oxidation according to the Anne method (Afnor X 31-109) or the Walkley–Black method (NF ISO 14235; GLOSOLAN-SOP-02). Organic carbon can also be measured by near-infrared spectrometry in accordance with ISO 17184:2014. Laser-Induced Breakdown Spectroscopy (LIBS) analyses can be performed on samples without preparation or directly in the field, but there are no associated standards. The choice of method depends on its suitability for the assessment objectives, the accessibility of the terrain, financial resources, available equipment and skills.

An overview of the available methods is provided below, along with a few examples distinguishing between methods used in the field and laboratory measurements on samples taken, as well as proxy-detection or remote sensing methods. Finally, many parameters can be estimated from basic measurements using Pedotransfer Functions (PTFs).

Field methods to measure soil properties

Methods that can be implemented in the field without scientific equipment can facilitate stakeholder participation in soil quality assessment, enable preliminary prospecting to inform a sampling plan for more in-depth analysis, and be integrated into conventional indicator measurements. These methods are the subject of numerous practical manuals and educational kits.

Some of these methods can be implemented with minimal impact on the soil, allowing repeated measurements to be taken in the same place over time. For instance, the Beerkan method (also known as BEST, or Beerkan Estimation of Soil Transfer parameters) can be used to observe water infiltration velocity on site, providing an estimate of saturated hydraulic conductivity that reflects soil porosity. One limitation of this method is the time required to acquire data, which can take up to two hours depending on the site.

Penetration resistance, which reveals compacted soils that restrict root growth, can be measured using a penetrometer. This method is straightforward to implement and can be used to cover large areas.

Other methods require the collection of a small soil sample. For example, structural stability can be assessed visually using a slake test (Herrick and Jones, 2002). For this test, aggregates are taken at two depths (0–2 cm and 2–10 cm) and a score is assigned based on how much they disintegrate or disperse in water over time.

Several field methods can be used to evaluate biological activity by measuring the rate at which organic matter degrades. The most common method involves burying litter bags (Lecerf, 2017). Weighing these containers before and after they have been in the soil for an average of three months provides an indication of biological activity.

Another method, the Bait Lamina method (governed by ISO 18311, 2015), uses thin PVC strips with 16 holes filled with an organic substrate (including cellulose and agar). Over time, soil organisms consume the organic matter in the soil, and the number of empty holes provides an indication of this biological activity. Biological activity can also be assessed by burying tea bags (the "Teabag Index"; Keuskamp *et al.*, 2013). However, caution should be exercised when interpreting the results obtained by observing the decomposition dynamics of standard materials, as these differ when it comes to natural litter.

Soil description and sample collection

Other approaches for characterising soil quality require physical access to soil characteristics, either through visualisation or by collecting soil samples that are subsequently destroyed. This means that the same measurement cannot be repeated over time.

These observations mainly concern morphological descriptions of soils, such as colour, signs of erosion, texture and structure. Some methods are applied to the entire soil profile, while others are only applied to the surface horizon.

Methods of describing soil profiles, which are usually linked to various soil classification systems, are detailed in reference documents such as the FAO Guidelines for Soil Description (FAO, 2006) and the French *Guide pour la description des sols* (Baize and Jabiol, 2012). These documents enable a detailed analysis of the soil structure, taking into account its spatial variability in both the vertical and horizontal planes for the cultivation profile (Gautronneau and Manichon, 1987). Observation allows for a precise diagnosis of the impact of crop protocols. However, implementation is time-consuming, involving digging and refilling the profile, and it requires considerable expertise, which generally limits its application to one repetition per plot. Recently, Tomis *et al.* (2019) proposed the 3D mini-profile method, which is based on sampling soil blocks of approximately one cubic metre using the blades of a telescopic loader or the bucket of a tractor. This method is popular with farmers as it enables real-time diagnosis in the field. Although easier and quicker to implement, it still requires considerable expertise (Boizard *et al.*, 2019).

To better account for all biologically derived structures (biostructures) observed in a soil profile, Piron *et al.* (2017) proposed an integrative method that quantifies the number of galleries (tubular structures) and the deposition of earthworm excreta in and on the soil surface, excreta being linked to soil aggregation. This method enables soil structure to be mapped across the entire profile at fine scale and rooting potential to be assessed. Simple methods are also available for sampling and counting earthworms, such as extraction using mustard or manual sorting of a block of soil. Ecological category identification (epigeic, anecic or endogeic worms) can be carried out visually in the field, but species identification and diversity assessment require expertise, which certain laboratories (e.g. Sol&Co, Auréa and OPVT at the University of Rennes) now offer.

The main tools used to describe the structure of the surface horizon are the Visual Evaluation of Soil Structure (VESS) method (Ball *et al.*, 2007; Guimarães *et al.*, 2011), whose interpretation grid is shown in Figure 5.4, and the Visual Soil Assessment (VSA) method (Shepherd, 2000). These methods are based on a block of soil extracted using a spade, hence the common name, "spade test". Methods based on the observation of blocks extracted with a spade produce indicators of structural quality that either explicitly or implicitly aggregate different parameters.

This type of assessment enables the soil structure to be characterised without making any value judgements. Indeed, a structural condition may be favourable for one function and unfavourable for another. This method can be used alongside other resources, such as conceptual diagrams illustrating the interactions between structural status, to formulate hypotheses and establish the origin of the observed conditions. Finally, it is possible to produce quantitative data by calculating the proportion of interesting conditions per horizon or per compartment. For example, this could be used to assess compaction.

Measurements on samples taken

Measurements taken from soil samples extracted from their natural environment can vary greatly depending on the properties targeted. Samples can be collected quickly and easily using an auger or from a pre-dug soil pit. The latter method cannot be used over large areas, but it allows samples with preserved structure to be collected. These samples can then be used to characterise the soil structure using tomography or to observe the microstructure using microscopy. Sampling in urban areas can be more difficult (due to stoniness and embankments) and often requires adjustments to be made, particularly regarding the sampling depth. In all cases, attention is paid to the location of the sampling point within the plot of interest, the type of sample (disturbed or undisturbed) and the conditions in which the samples are taken and stored. These conditions must be controlled to minimise errors in the final result.

Once the sample has been collected, performing measurements using analytical methods often requires specialised equipment and skills, which can limit accessibility. However, by using standards such as ISO or Afnor, these methods guarantee the validity and comparability of the obtained values and document the associated uncertainty. These basic measurements are fundamental not only for interpreting remote sensing signals, but also for calibrating other methods and for developing pedotransfer functions. In many cases, measuring an elementary quantity involves several operations and methods in different stages. These include preparation stages (impregnation, dissolution, cultivation, extraction techniques, reagent addition, etc.) and measurement stages (image characterisation, dosages, reaction dynamics, DNA analysis and organism identification, etc.).

To reflect the variety of processes that can be performed on soil samples in a laboratory setting, the following examples offer an overview of the most prevalent ones:

- dessiccation, combustion and dry or wet digestion, which allow the approximation of mineralisation rates, organic matter and nutrient content, for example;
- colour/fluorescence/spectrometry for identifying and measuring certain elements while taking into account their physical properties (e.g. nutrients and contaminants);

Figure 5.4. Extract from the VESS (*Visual Evaluation of Soil Structure*) analysis grid. According to Guimaraes *et al.*, 2011.

Structure quality	General appearance	Size	Roots	Visible porosity	Appearance after break-up: same soil different tillage		Distinguishing feature	Appearance and description of natural or reduced fragment of - 1.5 cm diameter	
Sq1 **Friable** Aggregates readily crumble with fingers	**No clods present**	Mostly < 0,6 cm aggregates	Roots throughout the soil. Roots present both within and around aggregates	Most aggregates are highly porous	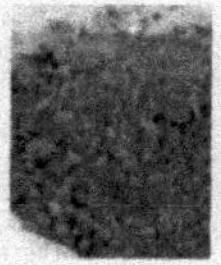	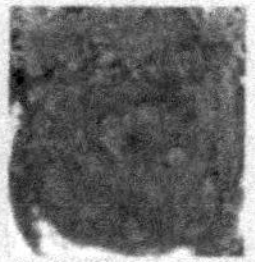	 Fine aggregates		Large aggregates are composed of smaller ones, held by roots. The action of breaking the block is enough to reveal them
Sq2 **Intact** Aggregates easy to break with one hand		A mixture of rounded aggregates from 2 mm to 7 cm		Most aggregates are porous	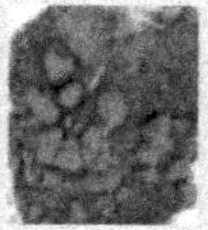	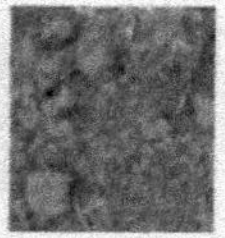	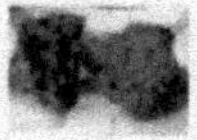 High aggregate porosity		Aggregates are rounded, fragile, porous, crumble easily
Sq3 **Firm** Most aggregates break with one hand	**Some angular, non-porous aggregates (clods) may be present**	A mixture of aggregates from 2 mm to 10 cm. Less than 30 % are < 1 cm		Macropores and cracks present	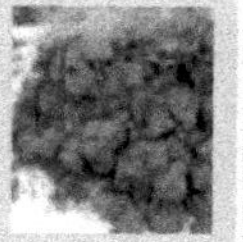	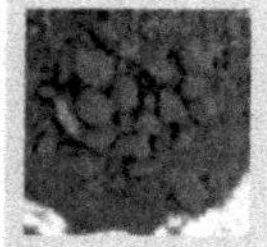	 Low aggregate porosity.		Aggregates have few visible pores and are rounded
Sq4 **Compact** Requires considerable effort to break aggregates with one hand	**Mostly sub-angular and non-porous aggregates**	Less than 30 % of aggregates are < 7 cm. Horizontal/platy structure also possible	Few roots, if any, and clustered around aggregates	Few macropores and cracks	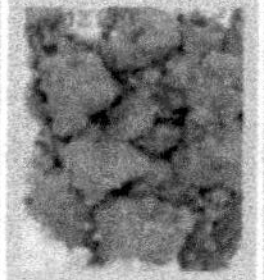	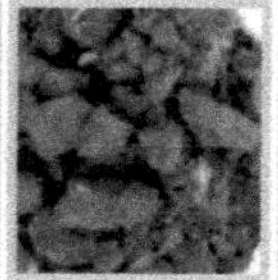	 Roots restricted to cracks and visible pores	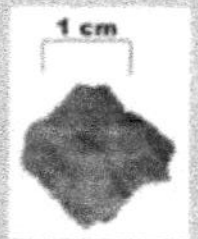	Fragments in cube shapes which are sharp-edged and show cracks internally. They are easy to obtain when soil is wet

• mercury porosimetry and gas adsorption to determine pore size distribution, stability indices and porosity (macro- and micro-), as well as the amount of available water/air;
• image analysis of 3D tomograms to characterise soil structure morphology and topology;
• fumigation-extraction for measuring microbial biomass;
• extraction and analysis of environmental DNA (known as "-omics" methods) to measure biomass, taxonomic richness, community structure, and microbial balance, or Phospholipid Fatty Acids (PLFAs) to estimate biomass and characterise the structure of the microbial community;
• extraction and identification of soil organisms at species or functional group level, enabling characterisation of the abundance and diversity of faunal communities;
• measurements of certain gas pressures (particularly oxygen) in relation to biological activity.

Non-destructive field approaches: proximal detection and remote sensing

Characteristics of measurements made using proximal detection and remote sensing

It is difficult to assess soil quality without digging a pit to access the different horizons and sampling their components. This involves destroying the soil locally, which limits the possibility of returning to the same place for further analysis and therefore does not facilitate dynamic monitoring of its properties. To overcome this limitation, extensive research has been conducted over the last few decades into non-destructive methods. These methods are based on proximal detection and/or remote sensing, and involve interpreting a physical signal whose characteristics vary according to the soil properties. Proximal sensing is carried out in contact with the soil, typically using electrodes.

In remote sensing, measurements are taken from a distance using instruments carried by drones, aircraft or satellites. Proximal sensing instruments rely on electrical conductivity and the reflection of electromagnetic waves emitted by radar, which are impacted by the characteristics of the environment. They also rely on the analysis of electromagnetic fields and, as with remote sensing, spectral or thermal imaging. These methods can be used to map soil indicators over areas ranging from a few square metres to several dozen hectares for proximal sensing and from a hundred square metres to the entire national territory for remote sensing.

The high temporal frequency of revisits in satellite images makes it possible to map spatio-temporal changes. Depending on the selected satellite configurations, these techniques can offer high spatial resolution, enabling homogeneous areas within a plot to be identified. Meanwhile, proximal sensing instruments are widely used at the intra-parcel level in agricultural environments for precision farming.

These tools take indirect measurements of one or more soil properties, which must be interpreted in relation to measurements taken on samples. This interpretation is based on models that link the detected signal to the value of the measured parameter.

If the proximal signal is spatially variable, it can also be used to guide sampling,[60] with a view of performing conventional measurements (which in turn feed back into the validation of the models).

Measurements of basic parameters and integrated measurements

The signal captured by proximal detection or remote sensing can be linked either to the measurement of a basic parameter, such as the water or carbon content of the surface horizon, or to a more integrated assessment of different quality-related dimensions.

Regarding basic measurements (Table 5.2), these techniques can be used to assess inherent properties such as clay content (by electrical resistivity), sand content (by magnetic susceptibility) and soil depth (by resistivity or ground-penetrating radar), as well as dynamic characteristics such as water content (by ground-penetrating radar), organic matter content (by infrared spectroscopy) and metal content (by magnetic susceptibility). These techniques are increasingly being used for the chemical characterisation of soils (e.g. near- and mid-infrared spectroscopy), and also appear to be relevant for understanding the Corg/Ntot ratio.

To date, most approaches that link the detected signal to basic soil quality parameters measured in this way involve validating and calibrating the associated models. However, these techniques have not yet been systematically evaluated in terms of the expected prediction accuracy according to the sensors, spatial and spectral resolutions, and the agroecosystems or ecosystems studied. In urban environments, the implementation of these techniques presents specific difficulties related to the condition of the soil surface.

Regarding complex indices, their relationship to a signal or combinations of proximal detection or remote sensing information is less often based on a validation approach. Notable exceptions include studies by Veum *et al.* (2017) and Adhikari *et al.* (2022), which assess soil quality or health indices using geophysical tools. However, no research currently map integrated soil quality or health information. This is a real challenge for current and future research dealing to the characterisation of spatiotemporal changes.

A great work has been done on carbon assessments. These combine process models that simulate carbon fluxes via crop cover using remote sensing data. However, these models still need to be developed in conjunction with spectral and/or spatial soil models. The challenge of verifying the effectiveness of carbon storage in soils to support incentive policies, such as the Carbon Farming project, has led to the development of certification systems that adopt a framework known as MRV (Monitoring, Reporting and Verification). The aim is to ensure that commitments made by stakeholders are monitored and reported to regulators, and then verified. However, various sources of uncertainty (spatial, instrumental, analytical and related to process model parameterisation) still need to be assessed.

60. See above, section "Sampling issues", p. 83.

Table 5.2. Feasibility of estimating basic soil properties using proximal or remote sensing.

Indicator	Proximity detection	Remote sensing
Nitrogen and carbon mineralisation		
Microbial biomass		
Carbon fractions	LAB	
Soil carbon stock (1 m)	LAB/FIELD	REG
Soil carbon stock (30 cm)	LAB/FIELD	REG
Organic carbon/clay ratio	LAB/FIELD/PLOT	PLOT/REG
Soil organic carbon content (g/kg)	LAB/FIELD/PLOT	PLOT/REG
Organic pollutants	LAB/FIELD/PLOT	PLOT/REG
Inorganic/metallic pollutants (TMMs)	LAB/FIELD/PLOT	PLOT/REG
Available potassium	LAB	PLOT/REG
Available phosphorus	LAB/FIELD/PLOT	PLOT/REG
Total nitrogen	LAB/FIELD/PLOT	PLOT/REG
CEC	LAB/ FIELD /PLOT	PLOT/REG
pH water	LAB/ FIELD /PLOT	PLOT/REG
Water retention capacity	REG	PLOT/REG
Available water capacity	FIELD /PLOT	PLOT/REG
Saturated hydraulic conductivity	LAB/ FIELD /PLOT	PLOT/REG
Electrical conductivity	LAB/ FIELD /PLOT	PLOT/REG
Structural stability	LAB/PLOT	
Roc fragments content	FIELD /PLOT	
Bulk density	LAB/ FIELD /PLOT	PLOT
Particle size distribution – texture	LAB/ FIELD /PLOT	PLOT/REG
Erosion rate	FIELD /PLOT	REG
Soil depth		REG

In green, direct estimate already published; in yellow, direct estimate still to be tested or indirect estimate (with other covariates) already published; in orange, *a priori* estimate impossible; in grey, estimate not yet proven. Level: LAB, spot measurement in laboratory; FIELD, spatial measurement in the field; PLOT, spatial measurement at intra-plot level; REG, estimate at regional or *supra-regional* level.

Limitations of proximal detection and remote sensing approaches

The main limitation of these techniques is that remote sensing signals primarily relate to the surface, or topsoil, and generally deteriorate in quality with increasing soil depth in the case of proximal sensing. It should also be noted that these tools have not yet been applied to the biological component of soils. In remote sensing, airborne devices are expensive, and acquiring data still presents many challenges, such as matching spatial and spectral resolutions, correcting atmospheric effects, dealing with cloud cover and shadows, and implementing strategies for selecting acquisition dates and temporal image mosaicking to account for the variability of vegetation development. Other issues relate to disruptive factors on the ground surface, such as humidity, roughness, the presence of plant residues or vegetation cover, and influences from other soil properties that exhibit similar spectral behaviour. In urban environments, the heterogeneity of surfaces poses particular challenges for the use of these techniques.

Finally, the work currently emerging in the field of acoustic detection can be considered as an extension of soil-based sensor techniques. At this stage, the focus is particularly on analysing macrofauna activity and soil structure dynamics.

Modelling approaches

Pedotransfer functions

The assessment methods described above are time-consuming and require specific equipment and technical expertise. Therefore, it may be useful to estimate the values of certain parameters using more accessible information and models of varying complexity, known as pedotransfer functions (PTFs). There is a wealth of scientific literature on PTFs, including models and pedotransfer classes, which provide values for different texture classes, for example. However, it is important to take the type of soil into account when using them, as estimates made for one group of soils cannot be transferred to other groups. Ideally, a PTF should be used in a pedoclimatic context similar to that in which it was developed, combining it with an assessment of uncertainty where possible and using recent measurements.

Table 5.3. Parameters frequently used in pedotransfer functions to calculate water indicators or bulk density.

Indicator	Water retention capacity (L/m²)	Saturated hydraulic conductivity	Bulk density
Bulk density (g/cm³)	×	×	
Texture	×	×	×
Abundance of rock fragments (%)	×	×	
Organic matter content (%)	×		×
Aggregate structure (class)		×	
Slope (°)	×		
pH			×
Cation exchange capacity			×
Sampling depth			×

PTFs are commonly used to estimate structural indicators, such as bulk density, and water parameters, such as available water capacity, water content at field capacity and wilting point, and saturated hydraulic conductivity (see Table 5.3).

Other types of modelling

Beyond PTFs, more complex modelling approaches are also being developed. With regard to bacterial richness, for example, a predictive model can be used to determine the values that correspond to a given set of soil and climate conditions. However, the current limitation of this model is that it does not consider interannual and/or intra-annual temporal variability (Terrat *et al.*, 2017). This model is based on the RMQS database containing 2,173 points of soil bacterial diversity, which has been interpolated to produce a map covering mainland France (see Figure 5.5). Similarly, Djemiel *et al.* (2024) have proposed a map of fungal richness for France based on the RMQS.

Recent developments in methods and their prospects

A review of the literature reveals that chemical and physical (including morphological) parameters are predominantly used to characterise soil quality. There is little questioning

Figure 5.5. Map of the taxonomic richness of soil bacteria in France (Karimi *et al.*, 2018).

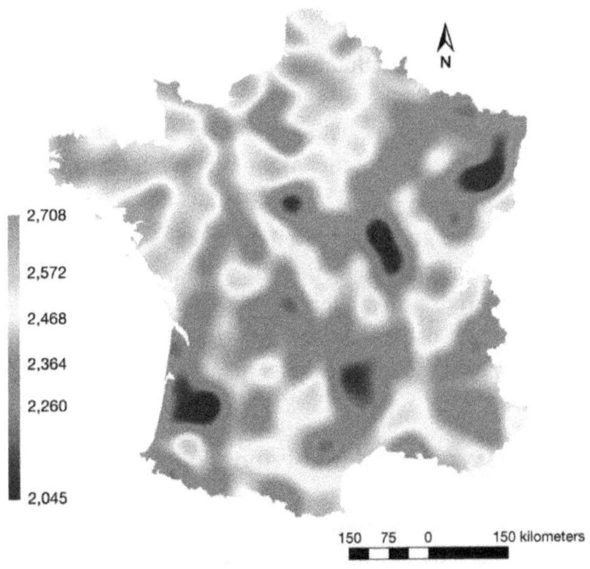

Bacterial diversity in number of TUs (taxonomic units per gram of soil).

of older parameters (those used before 1990), even though the methods used to measure or assess them may have evolved alongside technological advances. Conversely, the use of biological parameters to assess soil quality has improved considerably over the last 20 years (Bonilla-Bedoya *et al.*, 2023), primarily due to technological advances that have made information more accessible, as well as an increased recognition of the links between soil organisms and functions. This has led to a growing interest in soil biodiversity.

Although the parameters measured have changed little in recent decades, the methods used to sample, measure and estimate indicators have undergone significant development. Consensus on the method to be used is rarely reached, even for well-established parameters such as organic carbon content, pH and particle size distribution. These differences are often closely linked to the scientific discipline (e.g. for evaluating particle size) and/or the country (e.g. for evaluating pollutants), particularly within national measurement networks.

Therefore, there is a debate about the benefits and limitations of greater harmonisation and standardisation of measurement methods. Harmonisation is promoted on the one hand to ensure the comparability of measurements and the interoperability of databases. Harmonisation is also required for parameters associated with regulatory mechanisms, whether these involve incentives (e.g. carbon storage) or obligations (e.g. restoration or compensation). The criteria for triggering and monitoring such measures are legally enforceable, so their measurement or assessment must be as reproducible and systematic as possible. Conversely, adapting the method to consider the context of the soil being studied, the study objectives and the availability of local reference data makes the measured values more relevant and improves accuracy in line with the local specific constraints. The issue of harmonisation arises in remote sensing, for example, where one advantage is the ability to take measurements over large areas. However, these recent data cannot be linked to pre-existing historical series. This is the case with the characterisation of organic carbon content at the plot level, for example.

In terms of the biological component, "-omics" approaches are playing an increasingly important role in characterising soil biodiversity. The advent of metagenomics has greatly enhanced our understanding of biological diversity, particularly in terms of developing tools to characterise the abundance and diversity of microbes, bacteria and fungi. These tools have now been validated and are widely used to characterise soil microorganisms, which is not the case for soil fauna.

Reference framework and interpretation framework

Measuring a parameter does not, in itself, provide any indication of the phenomenon we wish to characterise or evaluate. For instance, a measurement of electrical conductivity is meaningless unless we know whether the value is low or high, normal or abnormal, or desirable or undesirable. The measurement must be interpreted using a reference system that associates a range of values with meaning.

Reference situation

If the purpose of the assessment is purely descriptive, the values that constitute the reference framework are usually derived from the statistical distribution of measured values, or from predictive models based on existing values. However, this distribution can vary greatly depending on soil and climate conditions, as well as the type of land use. For instance, the same level of carbon storage must be interpreted differently depending on whether the soil is forest, agricultural or urban; whether it is cultivated or not; and whether it is sandy, loamy, clayey, calcareous or non-calcareous.

If the purpose of the assessment is normative, i.e. if it produces a judgement based on the measured value, the interpretation reference framework will include threshold or target values. A threshold value defines a degraded state and is intended to trigger corrective or restoration measures, whereas a target value defines a desired state–an objective pursued by managers or public policies.

There are various ways of identifying a reference situation for establishing threshold and target values:

• **comparing the soil under study to an "undisturbed" situation.** This is often how the reference ecosystem for ecological restoration operations is defined. However, in reality, such a situation is often fictitious, particularly in European territories where human activities and changes to ecosystems have evolved together throughout history. The undisturbed situation used as a reference is therefore often one that has been reconstructed through modelling. This approach pays the most attention to the heritage value of the soil and preserves its original characteristics;

• **identifying ecological tipping points** at which soil functions significantly degrade or improve. Above these thresholds, which may be biotic (e.g. the disappearance of keystone or engineer species) or abiotic (e.g. changes in hydrological behaviour or pollution), the ecosystem's resilience is insufficient to restore it to its initial state or promote an ecological trajectory towards that state. In practice, such ecologically established thresholds are rarely reported in literature;

• **comparing the soil under study with a target for functions.** This target consists of either maximising the ensemble of functions (according to a weighting to be defined) or balancing them with each other. For functions defined by flows, the focus is on the dynamics of these flows, such as the rate of organic matter degradation, water infiltration, and additional carbon storage. However, this type of metric does not take into account soils whose functions are not very dynamic, yet which play a major role in water and climate regulation, such as peat bogs or permafrost (permanently frozen soils). These metrics are in fact more suited to biomass production;

• **comparing the soil under study with a target objective for services.** The nature and level of the expected functions are then determined in direct relation to the intended use. For instance, the presence of an impermeable (or poorly permeable) soil layer at shallow depths may benefit wetlands but hinder crop growth.

Threshold values may be set out in legal texts and/or professional recommendations, and will refer to standardised norms where applicable. It should be noted that, for certain parameters, the value ranges will only be meaningful above a base value that is considered natural. For instance, certain chemical compounds only constitute contamination when present in quantities exceeding the geochemical background level.

In practice, the existing values for a given context or territory are usually considered as the reference points and are even used for normative assessment purposes. In such cases, a value is deemed good if it is normal in comparison to all existing values for the same context. This shift from a descriptive to a normative use of statistical distribution is often implicit due to a lack of better alternatives and is rarely justified on scientific grounds. This problem goes beyond soil quality and is also discussed in the field of human health, for example. Indeed, using the existing situation to define good status is problematic when a significant proportion of soils are degraded, as this degradation is then automatically considered statistically normal.

Even in principle, the use of threshold values is sometimes debated. Thresholds, or tipping points, do not always accurately reflect processes that evolve gradually. Consequently, situations that are actually similar may be considered very different (e.g. degraded or non-degraded) if they fall on opposite sides of the threshold. This can result in attention and decisions being focused on situations that are considered degraded, while more cost-effective preservation measures are neglected. It should also be noted bearing this in mind that the analysed economic literature reveals a reluctance to use threshold values that impose significant limitations on decision-making support. In fact, the impact of variations in soil quality exists regardless of the distance from the threshold. Nevertheless, the presence of thresholds leads us to focus primarily on variations in the vicinity of the threshold (e.g. the transition from undegraded to degraded soil conditions).

Available ranges of values

The identification of threshold values remains a heterogeneous element in the literature. For some parameters, they are well established. For others, they remain highly variable either because the threshold is highly context-dependent, the associated databases are not yet comprehensive enough to cover different contexts, or the measurement methods under development are not yet advanced enough to identify stable thresholds. Sometimes, thresholds simply do not exist. These values can be applied at different levels of contextualisation. For example, they can be applied to all French soils, or by combining land use and pedoclimatic context.

To obtain existing values, soil information systems are regularly updated with new data on soil profiles, the chemical and physical properties of soils and, more recently, biological parameters. Threshold and target values are taken from legal texts or published scientific literature. Table A2 in the appendix summarises the existing values available in GIS Sol databases or in the literature for the indicators included in the list in Chapter 6 entitled "Selected generic indicators" (p. 110).

Scoring or normalising indicators

The interpretation framework discussed in the previous section gives meaning to the measured value of an indicator. However, to enable comparison or aggregation between different soil quality dimensions, and therefore between indicators included in different frameworks, indicator standardisation is required. The challenge lies in reporting the variability of different indicators on a common scale–typically between 0 and 1 (or 0 and 100%)–and enabling the same mathematical treatment to be applied to multiple parameters with different units. As this scale is directional, it is necessarily normative, with 1 (or 100%) representing the best outcome and 0 representing the worst. For instance, it is impossible to standardise descriptive elements such as soil types. Conversely, it is possible to standardise the abundance of a biological group (e.g. earthworms), which is considered as an indicator of quality.

The process of converting a measurement into a score, known as indicator normalisation or scoring, is characterised by a scoring function. This function can be monotonic and increasing when the quality score moves in the same direction as the measurement ("the more the better"), monotonic and decreasing when the quality score moves in the opposite direction ("the less the better"), or log-normal or Gaussian when the maximum score is achieved at an optimal value. These functions' parameters are based on the statistical distribution of existing values, or on the deviation from a model prediction, or on threshold and target values where these exist. The choice of these reference values is crucial and is determined by the operational objectives of the assessment.

Multi-criteria aggregation forming a soil quality index

The aggregation stage involves constructing a single numerical value, i.e. an index that integrates various dimensions and enables an overall assessment and monitoring of its evolution over time. Although this stage is optional, implementing it requires the values of the various dimensions included in the index to be standardised or scored in advance.[61]

Many disciplines (e.g. ecology, economics and land use planning) debate the very principle of such an approach. Aggregating several scores into a single score is beneficial because it saves the reader from having to process a lot of information and consider the questions that divergent results may raise, depending on the soil quality dimensions considered. However, this aggregated information reduces explanatory power, which ultimately complicates the identification of corrective measures. For instance, an aggregated index measuring soil fertility cannot identify the factors limiting this fertility. Non-aggregated information is required to determine whether the issue is related to

61. See previous section "Scoring or normalisation of indicators".

excess acidity, in which case liming may be a suitable corrective measure, or a lack of nutrients, in which case mineral or organic fertilisation or stimulation of mineralisation may be considered.

There is also no consensus on the aggregation function that links each indicator value to the aggregate index value, i.e. the weighting between the different dimensions of the aggregate. The choices made at this stage can significantly impact the value of the obtained index. Consequently, very different results can be obtained from the same dataset. However, the definition of the aggregation function is rarely justified on scientific grounds. More often than not, it is based on an arithmetic mean. Equal weighting of each dimension is usually adopted by default. In other cases, the weighting is adjusted according to environmental and social specifics of the study, or decided by land managers in light of their objectives. It may also be negotiated between stakeholders.

Another option for aggregating different dimensions without weighting them is the "one out, all out" or disqualifying criterion approach. Based on the principle of limiting factors, it considers that soil functions are interdependent and that all are affected when even one is not fulfilled. Similarly, soil is considered degraded when any form of degradation (e.g. erosion, compaction, salinisation, loss of organic carbon, loss of biodiversity, pollution, excess nutrients or artificialisation) exceeds the critical threshold. The European Soil Observatory uses this method to calculate the percentage of degraded soils (Figure 5.6). The advantage of this disqualifying criterion-based approach is that it focuses attention and efforts on the most vulnerable aspects. However, it has been criticised for potentially discouraging the introduction of new indicators into the assessment system, since increasing the number of indicators to be considered increases the likelihood of encountering a disqualifying criterion.

Figure 5.6. Soil degradation barometer established by EUSO.

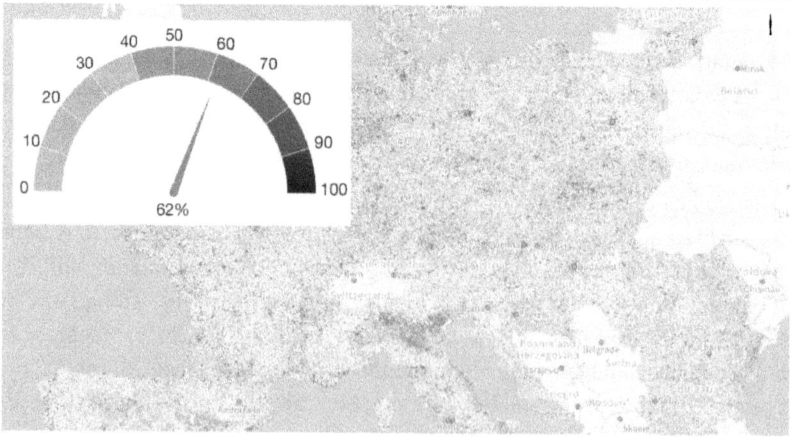

The barometer indicates the proportion of land affected by degradation processes. https://esdac.jrc.ec.europa.eu/esdacviewer/euso-dashboard (accessed on 4/11/2024).

The aggregation process can be applied at different levels. For example, it may involve aggregating the values of different indicators to score a function, or aggregating different indicators and/or functions to obtain an index representing overall soil quality.

Aggregating indicators to score a function

Concerning the function of "providing nutrients to the biocoenosis", one option is the enrichment index, which is based on free nematodes and provides information on nutrient dynamics, increasing with nutrient availability (Lu *et al.*, 2020). Another option is the Soil Biological Fertility Index (Renzi *et al.*, 2017), which aggregates soil organic matter, microbial biomass carbon, and various parameters of biological activity, such as basal respiration, cumulative respiration, and metabolic quotient.

Aggregation into an overall soil quality index

The Soil Management Assessment Framework (SMAF) (Andrews *et al.*, 2004) and the following Comprehensive Assessment of Soil Health (CASH), were initially developed for agricultural contexts in north-eastern USA and have since been expanded to other contexts. Today, the Cornell Soil Health Laboratory offers different indicator sets depending on the context and level of detail desired by the user, along with the associated aggregation methods.

The land use suitability index was developed as part of the UQUALISOL-ZU project (Keller *et al.*, 2012; Robert *et al.*, 2013). Combining land use versatility and local urban planning, it is based on six functions (water circulation and retention; nutrient retention and cycling; physical stability and support; biodiversity; filtration and buffering capacity; and soil heritage) and nine land uses from the CORINE LAND COVER nomenclature. For each function, a set of parameters to be measured (the optimum data set) is selected through expert assessment to characterise that function. Different land uses have different requirements in terms of soil parameters, so it is necessary to check whether each parameter has a threshold value according to its use. Depending on the result, the parameter is said to be adequate or inadequate for that use. Finally, as the index is based on the concept of limiting factors, if at least one of the parameters describing a function is inadequate, the function is deemed not to be satisfied, even if the other parameters have adequate values. The total number of satisfied functions for all uses can then be added up to create a land use versatility index. This provides an assessment of the land's suitability for each use identified in the study area.

The Soil Health Index (SHI) produced by Biofunctool® (Brauman and Thoumazeau, 2020), is based on nine indicators for which low-cost measurement methods can be implemented in the field. This enables the assessment of three main soil functions: carbon dynamics, nutrient cycling, and soil structure maintenance. To facilitate interpretation of the results by stakeholders in the agricultural sector, Biofunctool® provides a synthetic index ranging from 0 to 1, calculated using the aggregation method described by Obriot *et al.* (2016). Different levels of aggregation can be selected, such as the score per indicator or per function, or a total score which provides a multifunctional view of the system being assessed.

It should be noted that pollution is rarely included in comprehensive soil quality indices, although it is sometimes treated as a limiting factor for certain functions (UQUALISOL-ZU, MUSE). More often than not, pollution is the subject of a specific study in line with regulatory frameworks that are more closely associated with health risk management than the preservation of soil ecological functions.

Monitoring soil quality over time and space

Conventional soil mapping and digital soil mapping

Soil quality mapping aims to provide a geographical representation of soil properties. Like aggregation, mapping is an optional step in the assessment process. Soil quality can also be assessed and presented in statistical tables or diagrams. When mapping the results, the question of the surface unit at which the information is presented arises.

Conventional Soil Mapping (CSM) encompasses all approaches based on a model proposed by a soil scientist and verified by field observations. First, the mental soil-landscape model is created to define the initial soil contours based on landscape characteristics. This is done using interpretations of available geographical data (geology, existing soil maps, etc.), aerial photographs, and a representation of the relief. The soil scientist then selects the most informative sampling sites and optimises their spatial position to improve the quality of the information collected. The mental model can then be refined based on these field observations. Finally, the composition of the sampling unit used for mapping is determined. The map, which is considered a general model of soil classes, is supplemented with additional descriptions of soil profiles that characterise each mapping unit (Legros, 1996). Describing the soil using a shared classification system is therefore central to the CSM, as it specifically guides the division of the geographical area into soil units. From an operational point of view, the CSM is recognised as being quite time-consuming as it requires significant manpower.

Digital Soil Mapping (DSM), which is based on statistical modelling, has developed alongside digital tools and the increased availability of spatial data, such as digital elevation models and satellite imagery. Voltz *et al.* (2018 and 2020) define DSM as "the production of spatialised estimates of soil types or soil property values at any point in space using statistical models fed by spatial environmental data and calibrated with soil data available for the study area". Quality tests of DSM have shown that this approach often performs similarly to, and in some situations even better than, CSM. This is due, in particular, to the use of data mining approaches and the exploitation of soil information stored in databases, coupled with many available environmental covariates. CSM also offers interesting features, such as providing predictions of soil classes and/or properties that are quantitative and accompanied by uncertainty estimates, fine-resolution grid predictions (90 m) and a reproducible, quantitative spatial prediction model that allows products to be easily updated when new data becomes available. However, DSM approaches must be calibrated and/or validated using field data collected during CSM programmes.

Digital soil assessment

DSM can also form part of a broader Digital Soil Assessment (DSA) process. The aim is to translate the spatialised values of the measured parameters into an aggregate soil quality index. The data flow provides a single result based on a set of soil properties that characterise different soil horizons. Figure 5.7 illustrates the inference trajectories that can be used to produce a DSA result in three dimensions (a cube). Each dimension (or edge of the cube) corresponds to a type of inference:

- the scoring and normalisation[62] which translate the values of primary soil properties into a quality index;
- the combination of soil layers into a single value per spatial unit;
- the mapping by DSM.

The order in which these three processes are performed differs for each inference trajectory. Each trajectory produces a different result from the same initial dataset. In trajectory 1 (shown in red in Figure 5.7), for example, mapping is the final operation, performed after normalising the soil properties and combining the soil layers in the input dataset. This approach seems to be the most commonly implemented in recently published work. Trajectory 2 (shown in blue in Figure 5.7) begins by producing a map of each basic indicator/soil property. These indicators are then normalised into an index, and finally combined throughout the soil layers. Although recommended by Styc and Lagacherie (2019) for a regional assessment of the available water capacity, this approach does not appear to be used systematically.

Temporal monitoring

Soil characteristics vary according to different time scales. Those considered as inherent are characteristics whose associated variations can be observed over a period of several human generation, such as coarse particle content, slope and depth. In contrast, soil characteristics that are considered dynamic are sensitive to changes in the environment and management practices. These include organic matter, nutrients, pH and the abundance and diversity of soil organisms. Finally, some properties can vary significantly over a period of a few hours, including moisture content, the abundance of microbial populations and their associated activities (mineralisation, nitrification and respiration), as well as enzymatic activities. Therefore, it is essential to consider the time interval between two monitoring measurements in line with the temporal variability of the measured properties.

62. See above "Scoring or normalisation of indicators", p. 101.

This point is particularly relevant in ecology today, where the stability (and instability) of an ecosystem is one of the most widely studied characteristics, as an indicator of how ecosystems react to global changes. A substantial body of theory and methodology has therefore been developed for characterising the resistance and resilience of ecosystems in response to environmental or anthropogenic stimuli, as well as for identifying early warnings of change (de Bello *et al.*, 2021). Developing such a toolkit for soil quality change indicators could be highly effective in supporting conservation policies, for example by quantifying the degree to which soil degradation caused by artificialisation is reversible.[63]

Figure 5.7. Examples of numerical soil assessment trajectories following the sequencing of the stages of soil layer combination, normalisation (*scoring*) and soil mapping by statistical modelling (DSM).

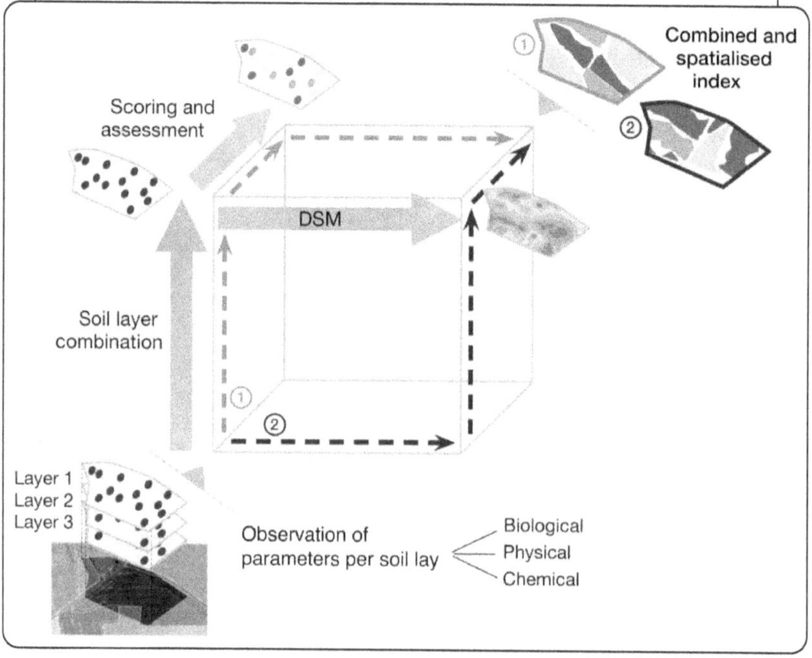

63. See Chapter 3, section "Regulating soil de-artificialisation and ecological restoration", p. 50.

6. Generic list of soil function indicators and assessment trial in a given area

Having broken down the process of implementing a soil quality assessment in the previous chapter, this chapter offers a preliminary selection of indicators based on their frequency of appearance in the literature corpus and their link to functions. This selection is accompanied by information on their inclusion in the annexes to the proposed European Soil Monitoring and Resilience Directive, as well as details of their operationality.

Main categories of indicators

The list of indicators for assessing soil quality is not a directory from which elements can be extracted and used independently; rather, it is an integrated whole. The indicators interact with each other and may have different statuses with regard to the assessment. Thus, the literature reveals a diversity of approaches to take such subsets into account, distinguishing categories of indicators that go beyond the classical attachment to the disciplinary field to which the measured dimension refers (physical, chemical or biological).

Role in relation to assessment

Indicators can be distinguished according to their role in the interpretation framework.[64] Some of them, not specifically related to soil and its functioning, are known as framing indicators, and define the general context of the indicators. This category includes at least indicators characterising the environmental context:

- soil type (based on a soil reference system such as the WRB or on basic parameters of texture, structure, pH water, organic matter content of the surface horizon, etc.);
- climate type (precipitation, temperature);
- topography (slope, altitude).

Depending on the purpose of the study,[65] the reference typology may also include land use and land cover (e.g., forest, urban area, orchard) and/or type of management (e.g., conventional or organic agriculture or soil conservation).

64. See Chapter 5, section "Reference framework and interpretation framework", p. 98.

65. See Chapter 5, section "Purpose of the assessment", p. 75.

Whether an indicator is positioned as a framing element or a dimension to be assessed in relation to this interpretation framework depends on its sensitivity to changes, the impact of which we seek to assess. For instance, if the objective is to evaluate the effects of management practices on soil compaction, the practices will be incorporated into the reference categories, with the soil structure serving as the indicator for evaluation. Conversely, if the objective is to assess the impact of compaction on soil microbial activity, the reference categories will consist of different degrees of compaction, and microbial activity will be the indicator considered for the assessment.

Specific indicators of functions and determinants

Regarding the link to functions, the information provided by the indicators may relate to soil characteristics that are either favourable or unfavourable to the function (known as "determinants"), or to an assessment of the function current performance. For instance, the function "regulating water quantity" can be characterised by the "hydraulic conductivity" indicator, which is related to density–an indicator of surface horizon structure. In this case, density is considered a determinant of the function "regulating water quantity". However, the concept of determinants is difficult to manipulate because of the strong interactions between ecological functions, meaning that many parameters are both determinants and indicators of functions.

Finally, the measurement of an indicator of function fulfilment may relate to different dimensions, such as content (e.g. contaminant or nutrient content), stock (e.g. organic carbon stock)–these two dimensions being linked by calculation, since stock can be obtained by multiplying the content by the bulk density, and then by the soil volume–, flux (e.g. infiltration or mineralisation rates) and state dynamics (e.g. change or maintenance of biodiversity indices, pH or cation exchange capacity).

Regarding biological indicators, the relationship to functions may differ depending on the variable measured. Distinctions are made between the abundance and diversity of a population or community–which can be broken down by level (individual, species, population, community) or type of biodiversity (genetic, compositional or functional)– and the activity of organisms (e.g. respiration, enzymatic activity, litter decomposition). Even within a given biological group, such as microorganisms, and despite all the aforementioned indicators being related to the function of "supporting biodiversity", different dimensions of the function are actually expressed when measuring:

- the microbial biomass, which broadly indicates the capacity to support biodiversity in terms of organism abundance;
- the taxonomic richness of bacteria and/or fungi, which indicates the capacity to support biodiversity in terms of the number of microbial groups and the biological functioning and resilience of the soil;
- the composition, e.g. the fungus/bacteria ratio, which indicates the capacity to decompose soil organic matter;
- the activity, such as microbial respiration, which indicates the capacity to transform organic and mineral matter, thus providing nutrients to the biocoenosis.

"Ideal" indicators

As functions are dynamic processes, a function indicator should theoretically reflect this characteristic, ideally quantify a flow or change of state rather than a state itself. For instance, the function "maintaining soil structure" should be represented by an indicator that reflects the dynamics of bulk density over time rather than bulk density itself or one of its determinants. However, the scientific literature does not provide much information about the actual link between the *indicandum* and the indicator when the *indicandum* is an ecological function. Consequently, an indicator is often identified based on its proximity to the fulfilment of a function rather than its ability to quantify this fulfilment.

To measure the performance of functions rather than the factors that determine them, a consultation was held with experts (three or four per function) to identify a series of "ideal" indicators based on the following principle:

- define a single indicator per function;
- choose the indicator that best represents the function;
- do not take into account the operational constraints on the implementation of the assessment of this indicator (the "ideal" nature of the indicator).

Depending on the situation, the selected indicator provides information on: the flux linked to the function (e.g. the annual amount of water drained at the base of the soil); the resulting change in state (e.g. the annual dynamics of pore size distribution resulting from the balance between pore creation, conservation, and disappearance); or the final state generated by the function (e.g. the structure of the food web). The resulting list (Table 6.1) aims to summarise each of the different functions and subfunctions defined in this study into a single value.[66]

Despite the attention paid to the link between each indicator and its associated function, the proposed definitions still reveal some limitations. For example, the amount of stabilised bioavailable contaminants does not specify which contaminants are considered, nor does it provide a precise definition of the concepts of bioavailability and the stabilisation of bioavailable contaminants (a process which is not irreversible). In other cases, the indicator does not reflect the full complexity of the object or phenomenon of interest. For instance, in the function "Providing nutrients", neither the annual amount of nitrogen assimilated by the biocoenosis nor the annual amount of nitrogen not leached takes phosphorus into account. However, phosphorus behaves differently in soils to nitrogen. The dynamics of the pore network are addressed solely from the perspective of pore size distribution, omitting connectivity and tortuosity. These factors are indirectly accounted for in a pore size distribution that varies over time. A food web could be characterised by the number or intensity of links, or by its stability or resilience, rather than by the number of nodes proposed here. Finally, these indicators are dependent on the chosen definition of each function they represent. Thus, while annual carbon storage across the entire depth could be a relevant indicator of the overall "storing carbon"

66. See Chapter 4, section "Six ecological functions that reflect soil quality", p. 63.

function, this option has not been retained given the breakdown into two sub-functions relating to carbon stock and additional storage capacity.

At this stage, these proposals constitute an initial approach to develop a system for indicating soil functionality, rather than a definitive, ready-to-use list. While these indicators, which focus on the level of function achieved, offer a fresh perspective on soil quality assessment, they do not allow us to diagnose the causes of the observed level of functionality or identify restoration strategies. In this sense, they complement the traditional soil condition assessment indicators, which are and will remain essential for understanding and managing the functions quantified using the "ideal" indicators, such as those proposed here.

Table 6.1. List of "ideal" indicators that correspond to the definitions of functions and sub-functions.

Function	Sub-function	"Ideal" indicator	Unit
Supporting biodiversity	Supporting vegetation	Potential net primary productivity	kgC/ha/year
	Supporting organisms	Food web (number of nodes)	
Storing carbon	C reservoir	C stock over total depth	kgC/ha
	Additional C storage	Additional C storage potential over total soil depth	kgC/ha
Providing nutrients to the biocoenosis		Annual quantity of nitrogen provided to the biocoenosis	kgN/ha/year
Regulating water quantitatively and qualitatively	Transfering water	Annual amount of water drained at the soil basis	m³/ha/year
	Storing water	Average annual amount of water stored over total	m³/ha/year
	Regulating water composition	Annual amount of mineral nitrogen not leached	kgN/ha/year
Regulating contaminants	Retaining contaminants	Annual storage of contaminants over total soil depth	kg/ha/year
	Reducing the bioavailability of contaminants	Annual amount of bioavailable contaminants which are stabilised	kg/ha/year
	Degrading contaminants	Annual quantity of organic contaminants-C respired	kgC/ha/year
Maintaining soil structure		Annual dynamics of pore size distribution over total soil depth	m³/ha/year

Selected generic indicators

Indicator selection strategy

The indicators in the list below were selected using a combination of three complementary approaches. The first approach consisted in identifying the soil properties that were most frequently recorded in scientific publications dealing with soil quality or health. Properties or parameters appearing in more than 20% of publications were selected. The second approach was based on the bibliographic corpus related to soil functions to identify the most relevant indicators for assessing the six functions defined in Chapter 4 (section entitled "Six ecological functions that reflect soil quality", p. 63). Cross-referencing these two approaches revealed a high degree of convergence between the two selection processes, which complemented each other to a small extent and justified the choice of a function-based approach to characterise soil quality/health. Finally, a third approach involved adding expert indicators to ensure the list proposed here was generic, i.e. relevant regardless of the context. For instance, although the earthworm-related indicators were derived from the selection process, enchytraeids were then added to address situations involving highly acidic soils. Such conditions are incompatible with the presence of earthworms; therefore, enchytraeids are used to assess processes to which earthworms contribute. Table 6.2 presents the selected indicators (around fifty). This list includes function indicators and the determinants of these indicators. Both are necessary for evaluating function.

Introducing the list of function indicators

Table 6.2 lists the indicators in the central column. Reading the table from top to bottom enables you to identify their relationships with the functions in the columns on the right, where a red box indicates a function indicator and a pink box indicates a function determinant. The indicators can thus relate to:

- a single function (e.g., saturated hydraulic conductivity for the function "Regulating water quantitatively" and microarthropod abundance for the function "Supporting soil organisms");
- several functions (e.g., soil basal respiration, which is an indicator of the two sub-functions related to supporting biodiversity);
- function determinants (e.g., bulk density, which indicates the function "Maintaining soil structure" and determines four other functions).

Reading the table from top to bottom allows you to identify the indicator(s) that characterise this function, or the determinants that contextualise it. For example, the function "Maintaining soil structure" is characterised by structural stability and bulk density, and is also determined by texture and organic carbon content.

Each indicator is accompanied by some informations related to the indication system. The area in which the measurement is to be taken is specified, whether that be on a

Table 6.2. Indicators selected to assess the soil functions identified in this study and indicators mentioned in the proposed *Soil Monitoring and Resilience Law* (SMLR) (version of May 2024).

Indicators

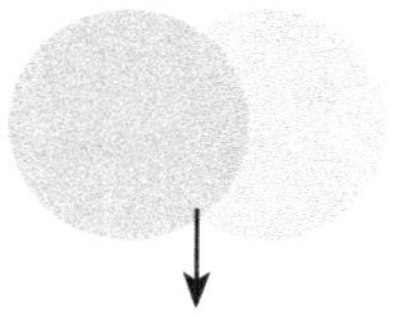

Indicator common to both appoaches

Threats mentioned in *SMLR***																							
Loss of biodiversity	Loss of soil organic carbon	Soil contamination	Excess nutrient content in soil	Acidification	Reduced water retention capacity	Subsoil compaction	Topsoil compaction	Salinisation	Erosion	Indicators	Indicator evaluation context	Standardised method	Operational maturity level	Use in economic evaluations	Supporting soil organisms	Supporting vegetation	Storing carbon	Regulating contaminants	Providing nutrients to the biocenosis	Regulating water quantity	Regulating water quality	Maintaining soil structure	
															Functions defined in this study								
										Depth	↕		●	x									
									■	Erosion rate	↔			x									
										Particle-size distribution -- Texture		x	●	x								■	
						■	■			Bulk density		x	●									■	
										Rock fragment content			○	x									
										Structural stability			○								■	■	
								■		Electrical conductivity		x	●										
					■	▒				Saturated hydraulic conductivity			●	x						■			
					■	▒	▒			Air capacity													
										Available Water Content	↕		●	x						■			
					■					Water Holding Capacity	↔												
				■						pH (water)		x	●	x									
										Cation exchange capacity (CEC)			●	x									
			■							Total N content		x	●	x					■				
			■							Available P content		x	●	x					■				
										Available K content		x	○	x					■				
		■								Total TMM* content			●					■					
										Extractable TMM* content			●					■					
		■								Organic polluants content (PAH°)		x	○					■					
		■								Organic pollutants content (PCB*, dioxins/furans)		x	○					■					

										Organic polluants content (pesticides and metabolites)		x	●									
										Organic carbon content		x	●									
										Organic carbon/Clay ratio			●									
										Carbon stock (over 30 cm)			●	x								
										Carbon stock (over 1 m)	↕		●									
										Carbon fractions			●									
										Oxidizable carbon fraction			●									
										C/N ratio			●									
										Humus type and composition (including C/N)			●									
*										Molecular microbial biomass		x	●									
*										Microbial biomass		x	●		*							
*										Soil basal respiration		x	●		*							
										N and C potential mineralisation rate			●		*							
*										Phospholipid fatty acids (PLFA)												
*										Soil bacterial diversity			●		*							
										Soil fungal diversity			●		*							
										Fungi/bacteria ratio			●		*							
										Enchytraeids abundance			●		*							
*										Earthworms abundance		x	●	x	*							
*										Taxonomic composition/functional diversity of earthworms		x	●		*							
*										Nematodes abundance		x	●		*							
*										Nematodes diversity		x	●		*							
*										Microarthropods abundance		x	●		*							
*										Taxonomic composition/functional diversity of microarthropods		x	●		*							
*										Enzymatic activity		x	●		*							
*										Abundance of ants												
*										Diversity of ants												

****** Version under discussion as of May 24, 2024 document 2023/0232 (COD) – 10236/24

*****TMM: trace metals and metalloids; PAH: polycycliv aromatic hydrocarbons; PCB: polychlorinated biphenyls

- (dark grey) Indicator for a function (right) or for a threat (in SMRL, left)
- (light grey) Optional indicator for a function or threat
- * One or more indicators can be selected from the column to characterise the function or state of degradation
- (white) Driver for a function or a threat
- (pale grey) Indicator not selected (in study or SMRL)

Indicator operational maturity:

- ● Mature indicator
- ● Maturing indicator
- ● Indicator under development

↕ Indicator assessed over the entire soil profile

↔ Indicator assessed on a catchment area and/or territory otherwise, the indicator is to be evaluated on a soil horizon specific horizon or soil layer, the entire soil profile, or at the scale of a catchment area and/or territory (i.e. a set of soil profiles). From a technical perspective, it is specified whether a measurement method has been standardised in the form of an ISO or Afnor standard. The operational level of the indicator (as set out below in "Operationality of indicators") and its use in economic context are also given. Please note that no mention is made of usage here: with the exception of the "Type and composition of forest humus" indicator, all the indicators listed here are intended for some assessment regardless of usage, particularly for territorial assessment or restoration operations.

Link to levels of degradation

The indicators listed in Table 6.2 are also compared with the soil threats (the ten columns on the left of the table). These indicators are mentioned in the proposal for a European directive on soil monitoring and resilience in its May 2024 version. There is relatively good agreement between the indicators selected for monitoring soil degradation in the proposed directive, and the function indicators selected from the scientific literature. However, of the forty or so function indicators selected, 14 were not included in the directive. Conversely, 6 of the indicators included in the proposed directive do not appear on the list of function-related indicators, primarily because they are highly specific or rarely used in assessments.

The dimensions considered in this study but not in the proposed directive in its May 2024 version are as follows:

• the available K content and the mineralisation potential of N and C are not included in the annex to the proposed directive, but are strongly linked to the functions, in particular "Providing nutrients to the biocoenosis" and "Supporting biodiversity";

• the type and composition of forest humus are also indicators strongly linked to the function "Supporting biodiversity", which is not included in the proposed directive.

The same applies to carbon fractions and the oxidisable carbon fraction. The latter make it possible to differentiate between the proportion of carbon that can be rapidly mineralised, and the proportion that is sequestered in the soil for longer periods. They are therefore essential for properly assessing the "storing carbon" function.

In the field of microbiology, the fungus/bacteria ratio selected for this study can be obtained using Phospholipid Fatty Acids (PLFAs), as proposed in the annex to the directive in its May 2024 version. While bacterial diversity is considered on both sides of the table, fungal diversity is only considered in the study. Although fungal diversity can be estimated using PLFA analysis, the resulting data provides a different level of detail to that obtained through high-throughput sequencing, which is based on soil metagenome extraction and target region amplification, allowing for taxonomic affiliation.

Conversely, ants are mentioned for their abundance and diversity in the proposed directive, but they are still uncommon indicators in assessments of functions. This reflects an openness to new bioindicators.

Relationships between indicators

The most common indicators used to assess soil functions are highly interdependent. It is important to consider these relationships when interpreting the results of a soil quality assessment. The list of indicators proposed in the "Selected generic indicators" section (p. 110) should therefore not be considered as individual pieces of information, but rather as a dynamic network of relationships. Therefore, when examining the values obtained for an indicator, it is necessary to consider how the parameters with which it is most closely related have evolved simultaneously. These related indicators are essential for interpreting the values obtained (e.g. depending on the pH level, the reference values for biodiversity indicators will differ considerably) and for ensuring an accurate understanding of the observed changes and the processes they reflect (e.g. a decrease in saturated hydraulic conductivity may be linked to soil compaction, as indicated by bulk density, and/or to the loss of organic carbon. These two issues do not call for the same management practices).

Operationality of indicators

The literature review carried out as part of this study shows that, although there is a broad scientific consensus on identifying the main soil quality indicators, it remains difficult to use these tools to inform soil quality preservation policies. These findings raise questions about the indicators operationality.

A framework to analyse the operationality of soil quality indicators

The use of indicators by users have to be based on a strategy that considers their specific needs (e.g. simple assessment or explanatory diagnosis, and the level of detail and degree of precision required), as well as the available resources in terms of time, skills, and tools. Therefore, operationality is not a measurable quantity, but rather the result of balancing information needs with the means of indication used.

This study assessed the operationality of each indicator selected in Table 6.2,[67] together with a measurement method (Table A3 in the Appendix), as well as a few so-called "ideal" indicators, as presented above. Operationality was considered in terms of its two dimensions: technical operationality (conceptual foundations, methods and database) and use-related operationality (time, skills and resources mobilised). These dimensions were broken down into different criteria drawn from the literature review. The aim is to differentiate between the elements that make indicators more or less operational, and to provide appropriate responses in terms of the resources deployed and the indication strategy.

The breakdown of operationality into different criteria is based on the CSLF approach presented in Chapter 5, in the section entitled "Selecting of a coherent and effective set of indicators" (p. 81). To this, a category relating to the diversity of implementation frameworks

67. See above, section "Indicator selection strategy", p. 110.

with which a given indicator is likely to be compatible has been added. In principle, the question of the cost of implementing a regular monitoring is included in the technical feasibility criterion. This criterion could be could be evaluated by adding equipment costs, time spent, and consumables used. However, the literature forming the corpus of this study does not include such information, which would have required a specific survey not carried out here. Additionally, the cost of implementation must be considered in relation to the value of the information produced, which is a developing research area in economics.

To provide an objective assessment of the level of achievement of all criteria of operationality, experts reviewed these criteria for each indicator. Common quantification methods were established and used by the experts, enabling a score to be assigned to each indicator. Depending on the criterion, standardisation may be binary (the criterion is either met or not met), gradual (operationality is defined in terms of tier, as shown in Box 6.1 and Figure 6.1), or combinatorial (a score value is associated with

Box 6.1. Combining approaches to operationality by TRL and by Tiers.

The technical operationality of a process or industrial development can be defined by prioritising the steps necessary to transition from the conceptual stage to implementation in real conditions and/or industrial production. Various authors in the reviewed literature have proposed two approaches for soil quality/health indicators: the TRL approach and the Tier approach.

The Technology Readiness Level (TRL) approach was developed by NASA in the United States and is now widely used across the industry. There are nine levels in total, with the highest being achieved when the technology is used in industry. When applied to soil quality indicators, TRL3 corresponds to the technical availability of the indicator. Promoting accessibility for analysis and interpretability for non-experts, as well as semi-industrial implementation at an acceptable cost, would enable TRL level 6 to be achieved. The existence of benchmarks, threshold values, or robust comparative approaches that enable diagnosis and provide management advice would lead to the highest TRL levels (8 to 9).

The Tier approach comprises three levels and is particularly prevalent in research areas focusing on impact assessment, such as ecosystem services, environmental impacts and greenhouse gas emissions. In the field of soils, it is used in various studies, not all of which interpret the three levels in the same way. Some approaches are similar to the TRL approach, classifying indicators according to methodological developments and the knowledge still required for their application. Others consider the Tiers based on the complexity of the indicators and methods used, ranging from indicators that are directly accessible to users to those that require complex modelling processes to be implemented. Although these two-Tier classifications emphasise different criteria, they both agree on categorising the degree of operationality of the indicators, with Level 1 being the most directly operational.

For the purposes of this study, these different scales have been combined into a single scale, as shown in Figure 6.1. We therefore consider maximum operationality at Tier 1, decreasing to Tier 3.

Figure 6.1. Proposed technical operationality scale, definition of its different levels and correspondences with the TRL and *Tier* scales.

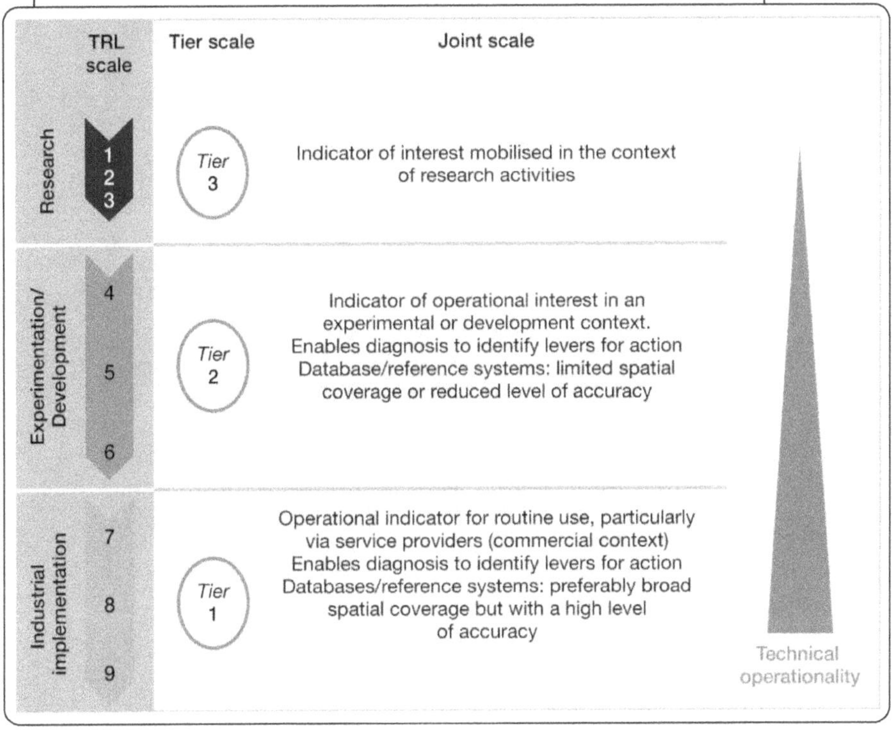

the number of situations in which operationality is ensured). Table A4 in the appendix details all the criteria and the quantification methods established for each one. Figure 6.2 summarises this information within the overall framework of technical and use-related operationality.

Operationality profiles and lessons learned

After a multiple factor analysis, the rating of each physical, chemical or biological indicator according to the various operational criteria reveals three main groups of indicators: indicators under development, maturing indicators, and mature indicators that are currently frequently used by stakeholders (see Figure 6.3 and Table A5 in the appendix). This gradient can be compared with the history of indicator use or data acquisition/aggregation efforts over the past 20 years. Physical and chemical laboratory indicators in the mature group have historically been used by multiple actors involved in soil management and gathered in various databases (see Box 6.2). Biological indicators, among the mature indicators, have been the subject of significant research efforts over the past 20 years, including the development of national databases and the active

Figure 6.2. Diagram describing the process for characterising the operationality of an indicator.

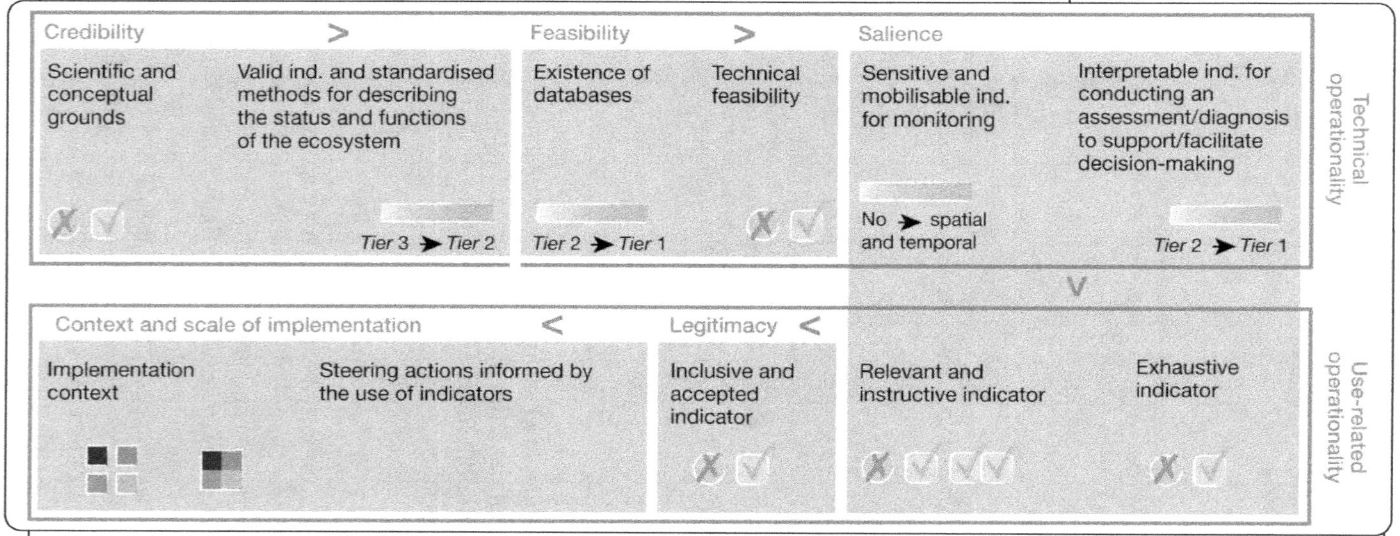

The symbols below each element represent the quantification method: binary (red cross: No; green tick: Yes; two green ticks: Yes – level 2), gradual (*Tier* gradient or amount of information), multiplicity of situations (mosaic) or implementation scales (plot or territory). Ind.: indicator.

dissemination of tools and knowledge to stakeholders through training, action research programmes and participatory research. Most of the indicators currently under development have a shorter history of measurement and dissemination, which is still ongoing, or are based on numerous small databases that need to be aggregated and consolidated at a national level.

Depending on the indicator, either the technical operability (e.g. measurement methods to be improved/standardised) or the use-related operationality (e.g. reference values for interpretation are incomplete, limiting their adoption by stakeholders) is not yet mature. For indicators under development, including the ideal indicators described above, priority should be given to stabilise measurement methods and collect data in accordance with these methods. Without this credibility, it is difficult to improve the other operationality criteria, even though these indicators are considered sensitive to management actions.

Figure 6.3. Availability of soil information at different scales for mainland France (Laroche *et al.*, 2024).

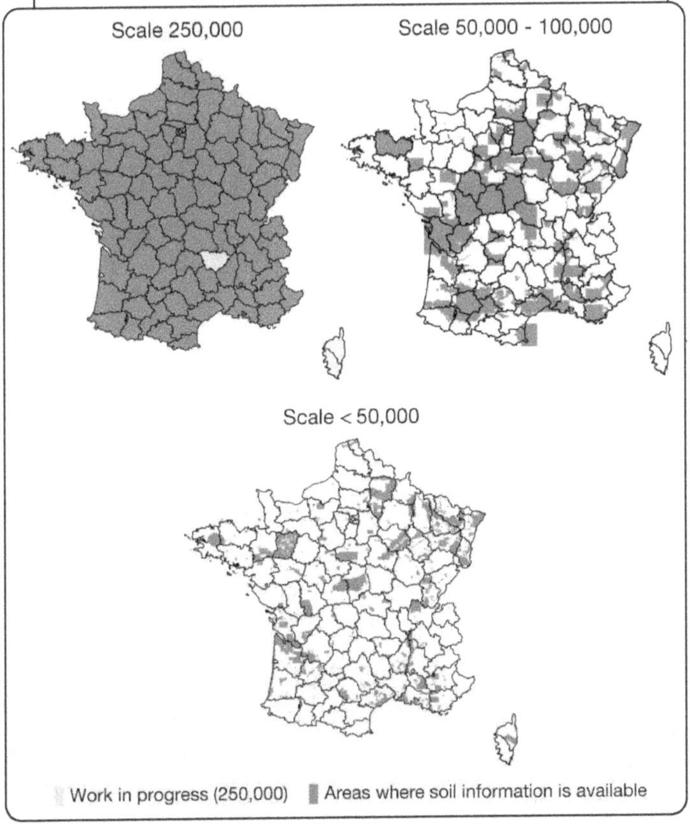

Table 6.3 Summary of indicator scores for the various operational criteria.

Score (100% → 0%)

Operationality

- **Credibility:** Scientific basis, Standardisation
- **Feasibility:** Technical feasibility, Databases
- **Salience:** Sensitive, Comprehensible, Comprehensive, Didactic
- **Legitimacy:** Accepted
- **Transferability:** Implementation, Steering

			Technical operationality								Use-related operationality		
			Credibility		Feasibility		Salience				Legitimacy	Transferability	
			Scientific basis	Standardisation	Technical feasibility	Databases	Sensitive	Comprehensible	Comprehensive	Didactic	Accepted	Implementation	Steering
Mature indicators	Physical	Bulk density											
		Particle size											
		pH in water											
		Available water capacity											
		Depth (m)											
		Electrical conductivity (Ece)											
		Electrical conductivity (Ec 1/5)											
		Rock fragments content											
	Chemical	Organic carbon content											
		Carbon stock (0-30 cm depth)											
		Total N content											
		Available P content											
		CEC: Cation exchange capacity											
		C/N: Carbon/Nitrogen ratio											
		TMMs – Pseudo-total concentrations											
		TMMs – Partial concentrations and agricultural context											
		TMMs – Partial concentrations and transfer											
	Biological	Microbial molecular biomass											
		Div_16S: Bacterial diversity (number of taxa)											
		Div_18S: Fungal diversity (number of taxa)											
		Microbial balance – Fungus/Bacteria ratio											
		Nematodes – Total abundance and functional groups											
		Nematodes – Nematofaunal indices											
		Earthworms – Abundance											
		Earthworms – Taxonomic richness and diversity											
		Lombrichians – Abundance of functional groups											
Indicators in maturation	Phys.	Structural stability – Weighted average diameter (Le Bissonnais method)											
		Type of forest humus											
	Chemical	Corg/clay ratio											
		C stock over total soil depth (up to 1 m)											
		Labile carbon/fractionation											

Table 6.3. (continued)

			Technical operationality						Use-related operationality				
			Credibility		Feasibility		Salience				Legitimacy	Transfera-bility	
			Scientific basis	Standardisation	Technical feasibility	Databases	Sensitive	Comprehensible	Comprehensive	Didactic	Accepted	Implementation	Steering
Indicators in maturation (continued)	Chemical (continued)	N and C mineralisation potential											
		TMMs – Partial concentrations and accessibility / health											
		Organic pollutants – PAH content											
		Organic pollutants – PCB, dioxin and furan content											
		Organic pollutants – Pesticide and metabolite content											
		Additional C storage potential over total soil depth											
		Available K content											
		Annual storage of contaminants / total soil depth											
	Biological	Microbial biomass – Amount of microbial carbon and nitrogen											
		Basal soil respiration measured in the laboratory											
		POxC: Labile carbon/fraction – permanganate-oxidisable fraction											
		Dehydrogenase enzyme activity											
		Beta-glucosidase enzyme activity											
		Phosphatase enzyme activity											
		Arylsulphatase enzyme activity											
		Enzymatic activity Ureases											
		Microarthropods and Mesofauna – Abundance											
		Microarthropods and Mesofauna – Taxonomic Richness and Diversity											
		Microarthropods and Mesofauna – Abundance of functional groups											
Indicators under development	Physical	Saturated hydraulic conductivity											
		Average annual amount of water stored / total soil depth											
		Annual amount of water drained at the base of the soil											
		Annual dynamics of pore size distribution / total soil depth											
	Chemical	Annual quantity of nitrogen provided to the biocoenosis											
		Annual quantity of mineral nitrogen not leached											
		Annual quantity of stabilised bioavailable contaminants											
	Biological	Potential net primary productivity											
		Food web – Structure											
		Enchytraedes – Abundance											
		Annual quantity of C inhaled from organic contaminants											

Box 6.2. The French soil information system.

Publicly available data for France is mainly stored in the National Soil Information System (Arrouays *et al.*, 2022), coordinated by GIS Sol and made available on its website*, as well as on the French open public data platform https://www.data.gouv.fr. The IGN also provides a wealth of geolocated data on the national land knowledge portal https://www.geoportail.gouv.fr.

The National Soil Information System includes several programmes: the Soil Inventory, Management and Conservation Programme (IGCS); the French Soil Monitoring Network (RMQS); the Soil Test Database (BDAT); the Trace Metal Database (BDETM); and the Urban Soil Database (BDSolU).

The IGCS programme is responsible for inventorying and mapping French soils at various scales:

- The Regional Soil Reference System (RRP) provides regional-level accuracy at a scale of 1/250,000 and currently covers 96% of French territory, although a significant proportion of the data on urban areas is unavailable (Laroche *et al.*, 2014). Across mainland France as a whole, it describes around 9,000 soil mapping units (SMUs) and around 17,000 soil typological units (STUs).
- The CPF section ("Connaissance Pédologique de la France", French program for soil mapping at medium scales (1/100,000 and 1/50,000), provides departmental-level accuracy, but only covers 18–24% of the territory (Richer-de-Forges *et al.*, 2014);
- The reference sectors (SR – "Secteurs de reference") at a scale of 1/10,000 provide detailed soil information at the level of small agricultural or environmental areas, but only cover a slight portion of French territory.

As shown in Figure 6.3, the degree of coverage of the territory is therefore uneven. It also varies depending on the land use: there is good knowledge of the quality of agricultural soils, less developed information on forest soils, and very incomplete information on urban soils.

The data produced as part of this IGCS programme show the distribution of soils according to the restitution scale. This graphical representation is linked to a DoneSol database (INRA InfoSol Unit, 2015).

The DoneSol database includes two sets:

- point data describing observations made in the field according to the DoneSol dictionary;
- area data, derived from the work of pedologists and mappers, describes the characteristics of the polygon(s) on the resulting soil map relative to environmental characteristics and soil types present in each area.

The RMQS is a systematic soil grid at a resolution of 16 km, enabling spatially homogeneous data collection using sampling methods that cover the entire territory.

The BDAT was created to consolidate and promote the over 250,000 soil analyses realised each year in France by accredited laboratories, primarily at the request of farmers. Soil data are physico-chemical characteristics from the surface horizon of cultivated areas, used to evaluate space and time soil variations.

The BDETM bring together the results of trace metal analyses carried out on soil samples taken from ploughed agricultural land throughout the country in connection with the application of urban sewage sludge. Since the 1990s, the BDETM has referenced more than 73,400 sites. Analyses carried out as part of scientific studies have also been added.

BDSolU database bring together data from various sources to discriminate natural pedogeochemical background values—primarily associated with the geological substrate—and values associated with human activities, which are particularly prevalent in urban areas and represent the anthropogenic pedogeochemical background.

* https://www.gissol.fr/le-gis/programmes/rmqs-34 (accessed on 9/11/2024).

Assessing soil quality in France

Data to assess soil quality

The spatial and temporal variability of soils, the diversity of parameters used to characterise their quality, evolving knowledge and a variety of characterisation methods result in a significant amount of soil data being produced. These developments enhance issues relating to storage, metadata management, data ownership and availability, and interoperability between information systems. These issues are crucial at all levels, from individual to international.

Most of the freely accessible data available for France is stored in the National Soil Information System (see Box 6.2). In addition, local data covering small areas may be stored elsewhere. Consulting enterprises specialising in the evaluation of one or more specific indicators (e.g. nematodes) also develop their own databases. The existence of such data is not centrally recorded, and access is not guaranteed.

The French GIS Sol then provides a centralised soil data management system and makes it possible to control the validity and comparability of soil data by standardising the measurement and calculation methods, particularly with reference to ISO standards. Most of the parameters measured as part of the French monitoring relate historically inherent soil parameters used to provide baseline values for assessing long-term trends. Recently, however, measurements of more dynamic characteristics have been incorporated into the RMQS for pesticide residue samples (see Table 6.1, p. 110 for organic pollutants) and microplastics. Assessments on the RMQS have provided estimates of existing values for biological indicators (e.g. enzyme activities, bacterial and fungal diversity, microbial biomass, and earthworm abundance and biodiversity) in different contexts.

The conservation of samples taken, carried out at the European Soil Sample Conservatory (CEES–Conservatoire Européen d'Echantillons de Sol) in Orléans, makes it possible to come back to soil samples and measure new soil parameters, depending on developments in concerns and knowledge.

Finally, interoperability between information systems enables soil parameters to be linked to other areas, such as the Nature and Landscape Information System (SINP–Système d'information sur la nature et les paysages) and forest ecosystems (RENECOFOR data).

However, these ongoing improvements incur significant costs and involve trade-offs based on prioritised needs.[68]

68. https://agriculture.gouv.fr/evaluation-du-groupement-dinteret-scientifique-sur-les-sols-gis-sol (accessed on 9/11/2024).

Examples of soil quality indicator assessments

The French Soil Information System supports the assessment process, particularly by providing existing values that form the basis of the reference framework for interpreting indicator values.[69] From this perspective, a test was carried out to determine the extent to which the currently available information can be used to assess the indicators set out in the proposed European Soil Monitoring and Resilience Directive, as well as a number of additional indicators. The test also aimed to specify some precautions to be taken when carrying out such assessments.

Assessing microbial indicators using RMQS data

Data on biological indicators, which are necessary for assessing soil ecological functions, are already available in France. Figure 6.4 illustrates the microbial biomass and diversity as determined using RMQS data in the Côte d'Or region compared to the national values. While microbial biomass is slightly higher on average in the Côte-d'Or department than the national mean, diversity is lower than the national median.

The distribution of the two parameters differs across the territory and varies between the two indicators: abundance is higher in the north-east, whereas microbial diversity is higher in the south-east.

Assessing the level of soil degradation, as defined in the proposal for the European Soil Monitoring and Resilience Directive

The proposal for the Directive includes the threshold(s) that describe one or more levels of degradation. Bulk density of deep horizons, for example, is an indicator of the level of deep compaction. Figure 6.5 shows the assessment of this indicator in the Côte-d'Or region, representing the proportion of the soil types considered as compacted in the study area. Very few soils are considered degraded in terms of deep compaction, except for a few in the north-west of the region.

Selecting a specific database to evaluate an indicator

Although there is a wealth of soil data (see Box 6.2, p. 122), it is not always easy to identify the data most likely to produce consistent results. Figure 6.6 shows an example of the assessment of the available water capacity, calculated using a pedotransfer function based on either the 1/250,000 RP or the 1/50,000 *Carte pédologique de France* (Soil Map of France). While the major groups appear to be correctly identified, the assessment at a given point may differ slightly, without either method necessarily being incorrect.

Therefore, the dataset must be selected according to the desired level of accuracy and the intended use of the indicator assessment. However, to simplify the process for users, Table 6.3 p. 120 summarises the specific assessments that can be carried out based on the available data.

---

69. See Chapter 5, section "Reference framework and interpretation framework", p. 98.

Figure 6.4. Microbial biomass (A; in µg of DNA/g of soil) and bacterial diversity (B; in number of OTUs/g of soil) in soils in the Côte-d'Or department (RMQS data): maps and comparisons with national RMQS data. OTU, Operational Taxonomic Unit.

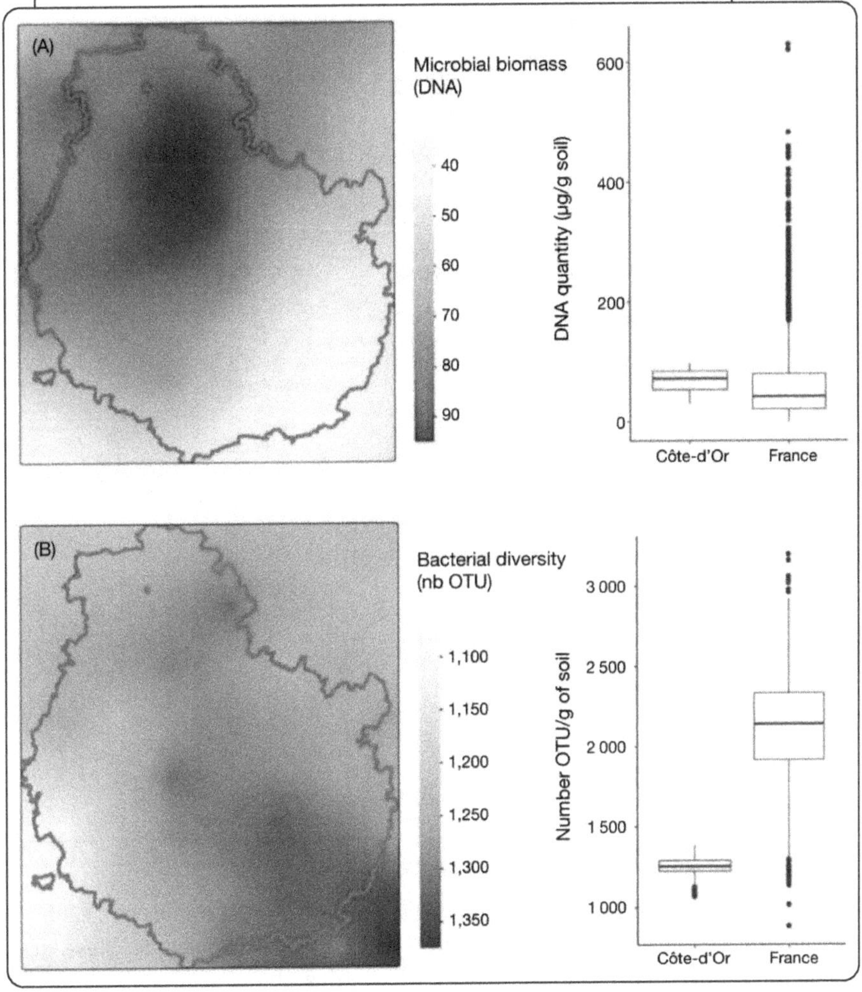

To go further

This study shows that the soil data currently available in France can be used to assess most of the indicators identified for characterising soil functions, quality and health. The main challenges now lie in the diversity of land uses, soil types and climates covered, and in repeating measurements over time. Increasing the frequency of monitoring measurements for certain indicators would enable the unprecedented environmental changes currently threatening soil quality to be better characterised and explained.

Figure 6.5. Available water capacity.
A) Assessment based on RRP data at 1/250,000 (depth = 110 cm).
B) Assessment based on CPF data at 1/100,000 (depth = 110 cm).
C) Assessment based on the *GlobalSoilMap* programme (depth = 200 cm).

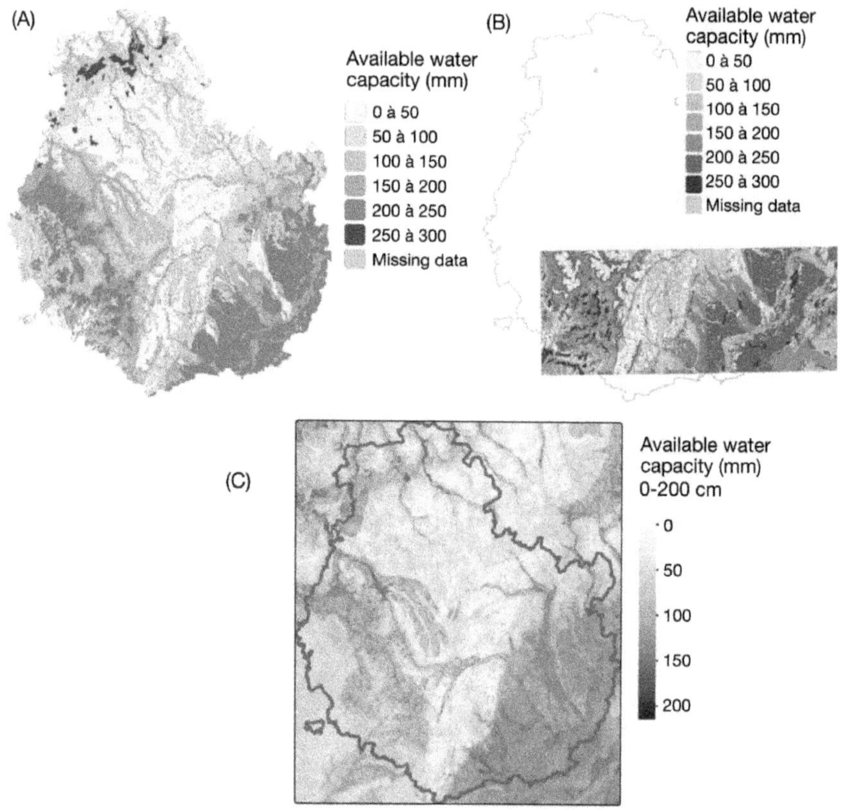

To this end, long-term field monitoring provides the basis for developing and calibrating new tools that enable data collection to be deployed across space and time. These include proximal and remote sensing tools,[70] which are increasingly being used alongside artificial intelligence processes. Participatory science and research also contributes to soil data collection[71] and covers a wide variety of contexts, including agriculture (e.g. OPVT),[72] urban and peri-urban areas (e.g. the REV-URBAIN project) and gardens (e.g. the JARDIBIODIV project).[73]

70. See Chapter 5, section "Advantages and limitations of certain measurement methods", p. 88.

71. See Chapter 2, section "Actors and mechanisms involved in soil quality assessment", p. 23.

72. https://ecobiosoil.univ-rennes1.fr/OPVT_accueil.php (accessed on 31/10/2024).

73. Participatory earthworm observatory, http://ephytia.inra.fr/fr/P/165/jardibiodiv (accessed on 31/10/2024).

Figure 6.6. Indication of the level of degradation due to compaction in the French Côte-d'Or department. Proportion (%) of each SMU whose STU exceed the density limit values proposed by the *Soil Monitoring Law.*

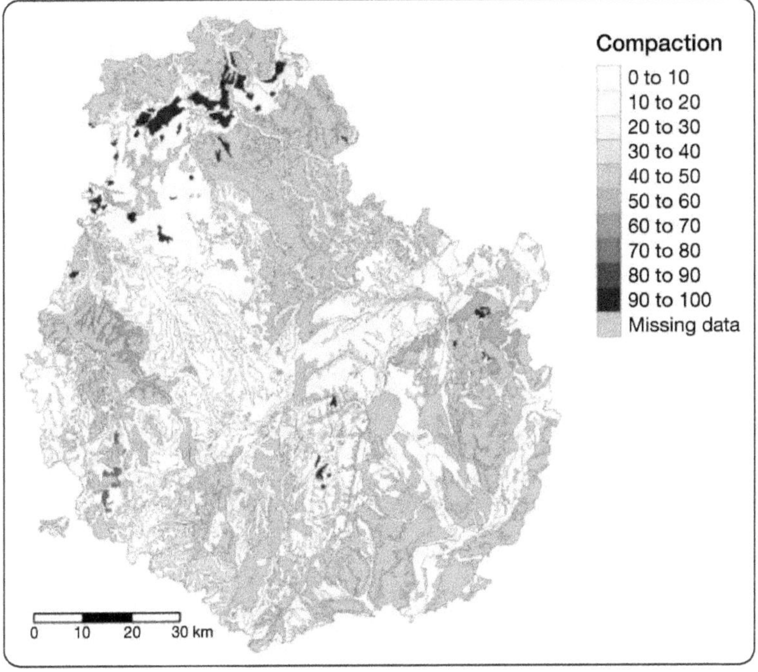

In the case of earthworms specifically, the OPVT has collected 6,820 observations over nine years, say an annual average of 757 observations. In contrast, research alone was only able to collect between 20 and 80 observations per year prior to 2010. This active participation has contributed to a database for the LANDWORM project,[74] which will soon provide fine-grained, contextualised reference values including diverse soil and climate conditions, and covering the entire French national territory.

74. https://www.fondationbiodiversite.fr/la-frb-en-action/programmes-et-projets/le-cesab/landworm (accessed on 31/10/2024).

Table 6.4 Availability of data regarding the main soil quality indicators.

Data source	Indicators provided for in the proposed Soil Monitoring Law												Other indicators					
	Electrical conductivity	Erosion rate	Organic carbon content	Corg/Clay ratio	Bulk density	Available P content	Total content of TMMs	Organic pollutant content (PAHs)	Available water capacity	Total N content	pH (water)	Microbial biomass	Depth	Granulometry – Texture	Cation exchange capacity (CEC)	Available K content	Carbon stock (30 cm)	Microbial density
IGCS (since 1990)																		
BDAT* (since 1990)																		
RMQS (since 2000)																		
Calculation by BDAT × IGCS*																		
Calculation by PTF from IGCS																		
Calculation by another method																		
Not available, to be measured																		

* Only on the surface horizon. **Green:** data available in each database (a light green colour indicates that data is available, but for a small proportion of the **database** points). **Orange:** indirect assessment based on available data. **Red:** information currently missing from French databases. BDAT: Soil Test Database; TMMs: Trace Metals and Metalloids; PTF: Pedotransfer Function; IGCS: Soil Inventory, Management and Conservation; K: Potassium; N: Nitrogen; P: Phosphorus; RMQS: French Soil Monitoring Network.

7. Lessons learned and avenues for research

Summary of key lessons

Expected soil qualities

Growing interest in soil ecology

The quality of something is assessed using a variety of approaches. Traditionally, it is defined as suitability for a particular use. Soil considered to be of good quality is therefore fertile for farmers and foresters, stable for builders, free from pollution for residents and rich in biodiversity for naturalists. However, this list reveals some contradictions. For instance, while the use of mineral fertilisers can maximise agricultural productivity, it can also potentially compromise water quality and soil organisms diversity. Similarly, while the drainage of wetlands can meet the objectives of agricultural productivity and disease vector control, it ultimately has harmful consequences for agriculture and human health by compromising the hydrological functions of soils. Growing knowledge of climate and biodiversity degradation, and their impact on the sustainability of human activities, has led to greater attention being paid to the ecological processes that maintain such qualities. The biological component of soil and the understanding of its interactions with physical and chemical components have thus become a subject of growing interest.

Distinctions between quality and health

The concept of soil health is gaining momentum, particularly within international frameworks that call for better soil conservation, such as those of the UN and the EU, and in international scientific literature over the last decade. This concept offers a more holistic view of the ecological functioning of soils than the traditional view of quality, which is more closely linked to their suitability for a particular use. Some studies also propose considering quality as the level of functionality that a given soil type can potentially allow (i.e. the commonly observed values for all soils of the same type) and health as the percentage adjustment to this level of functionality that is actually observed due to specific anthropogenic pressures exerted on that soil. Currently, however, the knowledge required to quantify such a ratio across all functions and according to different disturbance scenarios is lacking, given that the "undisturbed soil" reference situation is no longer available in reality for many soil types, particularly in France.

In search of a generic approach to quality: soil functions

A huge number of processes occur in soils and contribute to and regulate the major hydrobiogeochemical cycles that are essential for sustaining life on Earth.

These processes occur at different scales, covering the entire range of spatial and temporal dimensions. Examples include the rapid, highly local evolution of microbial communities and their metabolism, and the slower evolution of greenhouse gas levels in the atmosphere on a global scale. The concept of ecological function can be used to structure the analysis of these processes as a bundle of those that interact most with each other.

In this study, soil ecological functions were chosen as a means of assessing soil quality and health. This represents a compromise between an overly simplistic anthropocentric approach, which considers only certain aspects of soil quality, and an ecocentric approach that is less applicable to decision-making, since in ecology every object has ecosystemic relevance and there are no “good” or “bad” soil qualities.

Thus, the assessment of ecological functions is considered to be a common approach for all soil types and uses. The advantage of this generic approach is that it enables the impact of changes in land use, such as the conversion of natural areas for agricultural or urban use, to be measured or estimated and considered in land use decisions.

This function-based approach then complements the focus on the heritage value of soil naturalness. Indeed, assessing soil quality solely on the basis of functionality considers natural soils whose functionalities have been preserved on the same level as engineered soils that show the same functionalities.

Finally, an emerging perspective inspired by the ecological concept of functional rarity is proposed. At the scale of a given territory, this would involve considering soil units with a distinctive and scarce functional profile compared to the territory’s soils as a whole.

Soil quality, an often-overlooked aspect of land

After decades of focusing solely on soil productivity, the various other functions performed by soils and the benefits these bring to society are finally being recognised. In agriculture, specific practices are being promoted to preserve soil functions, such as reducing tillage, integrating grassland into crop rotations, and planting trees in fields. In urban planning, limiting the alteration of soil functions is now enshrined in law. Finally, initiatives that link scientific resources with citizen involvement have helped raise awareness of soil-related issues in recent years.

However, the economic value of land only considers soil quality in very specific cases, such as certain AOC areas. Land-related criteria, particularly proximity to urban areas, infrastructure and economic sectors, take precedence, as does agronomic value in terms of more or less highly valued crops (orchards, vineyards, cereals, grassland, etc.).

Currently, urban planning procedures pay little attention to the ecological quality of soil, except in the voluntary initiatives of certain local authorities associated with research projects (MUSE, UQUALISOL-ZU, etc.). Even when the law seeks to consider soil quality, as in the context of land development operations, its consideration proves to be of little decisive importance compared to socio-spatial issues surrounding land use in the final decisions. The maintenance of soil functionality by the administration is evidenced only through the application of some specific legal regimes, such as climate change mitigation, preservation of artefacts, and water potability.

The indication system: a stated purpose, indicators and a reference framework for interpretation

Prior to interpreting each indicator, an appropriate strategy must be developed, based on the purpose of the assessment, particularly with regard to the complementarity of the used indicators, their possible weighting, the soil sampling and the statistical and/or cartographic processing of the results. The combination of all these elements constitutes the indicator system.

This study outlines the stages of the assessment process at which the user must make choices. These choices form the basis on which scientific expertise can determine the relevant methodological options.

Clarifying the purpose first and foremost

Although it is often overlooked, it is essential to clarify the purpose of the assessment in order to select relevant indicators and determine how to take into account their spatial and temporal variability, as well as soil types, climates and land uses. For instance, an assessment aimed at monitoring the functions of all soils in a territory over time would be conducted very differently from a local study intended to select a location for an industrial project or a local study intended to identify the impact of a site's management methods on its ecological functions in order to make adjustments. However, in the scientific literature, the choice of indicators and the definition of how they are interpreted are rarely explained in relation to the purpose. Studies are most often conducted in a specific field of application (e.g. agriculture, forestry or urban planning) based on standard practices, without linking them to a theoretical or methodological framework.

Consolidated grounds for most indicators

The list of basic parameters involved in characterising and assessing soil quality is well established in the literature, allowing for the identification of around fifty main indicators. Given the continuous improvement in the collection and availability of data on these indicators, an initial knowledge base can be used to assess the physical, chemical and biological dimensions of soil quality in France. To further this assessment by using an approach based on the soil ecological functions, indicators at different maturity levels have been identified, especially those concerning biological indicators (see below, "Areas for research").

Active debates regarding measurement or estimation methods

The degree to which methods are harmonised is a recurring topic of discussion. This is an important issue because differences in values may be more due to different measurement methods than to different realities. Standardisation based on norms is therefore recommended with regard to measurement methods, in order to avoid misinterpretation. Discussions therefore remain ongoing regarding the most relevant methodological elements or steps to be standardised. The aim is to obtain standardised, comparable results while ensuring that scientific protocols remain flexible enough to suit the specific objectives and characteristics of each study.

The degree to which the produced information is aggregated is also a matter of differing opinion. Results expressed in the form of various indicators that evolve according to contrasting or contradictory dynamics can make it difficult for non-expert users to draw conclusions. Thus, within broad governance and communication frameworks such as the UN, the EU and national objectives relating to artificialisation, there are regular calls for a single index to be developed to communicate the overall evolution of soil quality. However, reaching a consensus on the resulting proposals is difficult (e.g. EU-level soil degradation barometer, or net artificialisation in France). These proposals risk overemphasising certain aspects of quality or offsetting aspects that evolve in opposite directions, thereby obscuring certain types of degradation. Aggregated information must therefore be interpreted by using the basic information in order to properly explain trends and identify improvement strategies.

The issue of uncertainty is also crucial for interpreting assessment results meaningfully. There are methods for quantifying uncertainties related to measurement and calculation methods, and progress in this area should enable them to be explained more systematically in scientific publications. This will enable these uncertainties to be addressed more effectively during aggregation and spatialisation operations.

Finally, concepts and models that establish links between ecological functions and their indicators are rarely developed. The choice of indicators is then usually made on an empirical basis.

Gaps in the framework for interpreting indicators

The values obtained for an indicator only have meaning in relation to the associated interpretation framework. This framework is established by taking into account the intended purpose of the assessment and the soil and climate context. The first step in the assessment process is to characterise the quality of the soil in question using a descriptive approach based on existing values. If the purpose is normative, i.e. if the aim is to assess the soil quality in order to decide on preservation or restoration measures, the indicator values will be compared with threshold and/or target values.

Existing values, thresholds and/or targets serve as the reference values. These depend on the reference situation against which the current state of the soil is assessed. The different methods for selecting the reference situation have been summarised and are the subject of unresolved questions:

- **based on a policy objective** (e.g. limiting the risks to the population from exposure to pollutants). The values may then differ depending on the country or region;

- **starting from a management objective** (e.g., a production objective for a farmer or forester), bearing in mind the bias towards considering production yield alone as an indicator of soil quality. This approach does not enable us to distinguish between the yield attributable to soil quality and the yield attributable to human interventions, such as the use of inputs. In this example, higher-quality soil may produce lower yields per hectare if managed to minimise inputs and interventions;
- **targeting the condition of undisturbed ("natural") soil of the same type.** However, such a situation is often fictitious in reality, particularly in European territories, given the extent to which human activities and ecosystem changes have combined throughout history. The undisturbed situation used as a reference is therefore often a situation that has been reconstructed through modelling;
- **scientifically identifying the thresholds for change in the state of the "soil ecosystem".** In a functional approach, this involves identifying the ecological tipping points at which these functions significantly degrade or improve. This area of research is still very much open, in connection with ecological work on ecosystem dynamics.

In practice, most of the thresholds mentioned in the scientific literature are derived from the statistical distribution of existing values. Thanks to the ongoing consolidation of databases, the identification of threshold values is gradually improving. Soil is therefore considered to be in "good health" when it falls within the normal range of values observed for comparable soils. This assessment strategy based on existing values is used in the absence of a better alternative, but caution should be exercised when interpreting results, as it does not take into account the level of degradation or regeneration of the reference soils. For example, if the reference soils have already been degraded, the soil under study will be considered to be in satisfactory condition even if it has also been degraded, and vice versa for a set of improved soils.

The question arises differently when assessing the effects of management practices that have been implemented over time in a given situation. In this case, the reference point is clearly the initial state of the soil at the start of the study. The question then becomes whether it is possible to discern the effects of the management practices from changes linked to factors such as climate change.

Data: a key aspect of operationality

One of the major obstacles to the operational implementation of the indication system is the limited availability of reference data for characterising soil quality. Some indicators do not have threshold values because these depend on usage and context, and existing databases are not consolidated at a sufficiently detailed level to establish them. To date, the proposed threshold values are predominantly global. France has an exemplary information system compared to other countries (see Box 6.2). More specifically, existing values can be found in specific databases or scientific articles. However, given the diversity of soils, climates and land uses found across the country, databases still need to be improved to adequately cover this diversity, both spatially and temporally. Little data is available specifically on urban soils.

Furthermore, to ensure that the French system for collecting and managing soil data contributes fully to the operationality of the indicators, users must have free access to all these data.

Towards an integrated scheme for preserving soil quality

A definition of quality to enhance legal certainty

Legal studies have highlighted the absence of a legal definition of soil quality, as well as the issues of legal clarity and certainty that this raises. Consequently, soil quality is treated in a paradoxical manner within governance instruments: although it is recognised as a prerequisite for life and human activity, it remains undefined and immeasurable, and is therefore largely absent from regulatory frameworks.

Gradually, environmental and urban planning laws are incorporating concepts from ecology and the environmental sciences. This is evident in the concept of functions, for example. In particular, legal literature, such as French and European positive law, refers directly to good ecological status in terms of air and water quality, but not yet in terms of soil quality.

However, even though the criteria for legally defining good soil quality have not yet been established, it is already possible to consider soil characteristics when developing measures to preserve this quality.

Insufficient hindsight for ecological soil restoration

Soils have a capacity for resilience linked to their ecological functions. However, once certain thresholds are exceeded, the damage becomes irreversible and ecological restoration operations are required. The main feature that distinguishes soils in the field of ecological restoration is that they evolve much more slowly than more directly observable ecological successions. Therefore, the effective restoration of soil functions may take longer than it takes for certain plant communities to re-establish themselves. Legislation and public policy may not keep pace with the requirements for the effective restoration of soil functions. Soil quality indicators should therefore be introduced to assess progress effectively.

However, soil quality indicators alone cannot be used to assess ecological restoration. The monitoring protocol must be supplemented with ecological indicators for the entire ecosystem, as well as economic, sociological, and cultural indicators. These indicators are crucial for ensuring the sustainability of such operations, which can be achieved through contractual mechanisms or land management practices that consider soil health.

So far, the scientific studies documenting such trajectories have been conducted over periods that do not allow conclusions to be drawn about the effectiveness of the ecological restoration of the studied soils. Consequently, there is a lack of insight into soil engineering operations, which calls for caution. Following the Avoidance-Reduction-Compensation (ARC) mitigation sequence, priority should be given to the recycling of degraded areas rather than restoration strategies implemented to compensate for new degradation.

Calibrating incentive schemes

In law, an indicator can take several forms and is monitored by various public and private bodies, including state and local authorities. Should soil health be enshrined as a legally binding objective tomorrow, it would be up to the legislator to decide whether to establish a special administrative policing regime or incentive mechanisms in the form of contracts or certifications. The two approaches are not mutually exclusive, but refer to different ways of ensuring compliance.

Economic literature focuses on the proportional calibration of incentive instruments, which is justified by the fact that those responsible for soil degradation often do not suffer the most from its consequences. These consequences can be felt at various scales (e.g. mudslides resulting from erosion-promoting practices, flooding resulting from loss of infiltration, or climate impacts such as warming resulting from greenhouse gas emissions), as well as over time (e.g. buying a previously degraded plot of land or bearing the future consequences of present degradation). Regulations on soil quality will be all the more justified if their economic consequences are precisely understood, and all the more effective if based on reliable environmental and economic data.

Information on soil quality is therefore crucial for both direct soil users and those whose living environment is changing in a given area.

Gaps in knowledge about the consequences of land use and practices

The literature emphasises the importance of basing incentives on obligations of means (uses and practices), provided that the scientific basis on which they are defined is improved and updated, and that the overall consistency of the implemented measures is ensured. With a few exceptions, incentives to preserve soil quality are based on empirical knowledge and ad hoc experiments. The results of these experiments are then extrapolated to designate certain uses and practices as beneficial to soil quality. However, the actual results obtained are rarely verified. In some cases, there is insufficient scientific knowledge to assess the real impact (e.g. carbon storage). Therefore, improvements are needed in terms of both metrics for some indicators and understanding the link between the implemented practice and the observed result in the field, taking into account other parameters (e.g. climate change).

Importance of stakeholders' involvement

A link has now been established between soil quality and its impact on quality of life, whether negative, such as flooding, or positive, such as urban green spaces and landscape diversity. This is a significant and tangible way to raise awareness.

The indication system creates a space for dialogue, aiming to involve all soil users in developing knowledge and action. It is therefore crucial that its implementation is not limited to sectoral public policies, but rather serves as an integrative tool for issues such as food, urbanisation and climate change. Another step in this process of appropriation is the collective prioritisation of soil functions.

Given the complexity of soil functioning and the issues raised by the definition of soil health, the scientific literature emphasises the importance of participatory approaches. Involving stakeholders in the design, implementation and interpretation of monitoring promotes a shared perception of soil quality and health, and helps to regulate power relations. The development of affordable field kits and educational manuals, and the establishment of facilitation structures such as living labs, are expected to create opportunities for involvement and discussion about soil, its uses, and its health.

Areas for research

Deepening the functional approach

An approach to soil quality based on its functional performance is well suited to collaborative efforts to preserve soil quality. It highlights the suitability of soil use and the trade-offs between functions in order to inform decision-making. However, this approach is less well suited to monitoring a territory, as this involves overcoming difficulties such as taking different functions into account in a single result and the fact that the link between indicators and functions is often very difficult to model and is usually based on empirical evidence. It is also difficult to distinguish between indicators that correspond to factors determining the fulfilment of a function and those measuring fulfilment directly.

However, like functional ecology, functional pedology appears to be a field of research that could provide answers to these issues, but much remains to be done. Better documenting the relationship between the fulfilment of functions, which refers to still-immature indicators, and determinants, which are easier to measure, therefore requires:

- the definition and selection of indicators which give information about the fulfiment of a function, for which this study provides a working basis;
- the development of a network for measuring soil functionality. This network would not be intended to cover all possible situations in an unbiased manner, but rather to serve as a learning and validation dataset. This could be based, in particular, on existing long-term experimental networks in which a number of functions are already being monitored. The issue would then be to consolidate them;
- the development of modelling strategies and their calibration based on measured situations (previous point);
- the extension of modelling strategies to all geographical points where the relevant determinants are monitored, or even to the whole of France, based on digital soil maps, for example.

Balancing the scientific corpus, which is dominated by agricultural issues

The scientific literature on soil quality overwhelmingly focuses on agricultural issues. Consequently, the selection of indicators and calibration of reference values are often suggested to address these issues. The consequences of degradation are most directly perceived by stakeholders in the agricultural sector, particularly with regard to the dynamic properties most sensitive to soil management practices. However, knowledge of the long-term consequences of the generalisation and intensification of anthropogenic pressures on soils shows the need to broaden the scope of study. For instance, low-productivity areas such as moors and wetlands fulfil vital hydrological, ecological, and climatic functions. Urban soils are mainly considered in terms of the risks to human health associated with pollution, flooding, and soil instability, rather than their ecological functions. Forest soils are the subject of growing concern regarding issues such as erosion and compaction, and specific tools adapted to the difficulty of accessing these areas are required.

Developing biological indicators

Although a set of biological indicators has existed for 10 years and has been more or less consolidated, it should be noted that some of these indicators are still not being implemented. Nevertheless, soil organisms and vegetation play an important role through the complex network of interactions that exists between them and the surrounding environment (abiotic), contributing to or affecting a multitude of services related to soil functioning. Therefore, it seems necessary to expand the range of bio-indicators used in soil health assessments. Certain organisms that are rarely used in soil health diagnostics represent a significant proportion of the edaphon biomass such as protozoa,[75] or play a specific role in soil functioning, such as enchytraeids or ants. It is also important to consider organisms that alter functions. For example, certain phytophagous insect larvae that consume living roots (e.g. wireworm or cockchafer larvae) will have a deleterious effect on vegetation development if present in excessive numbers.

Finally, beyond the indicator itself, new metrics need to be developed to capture the complexity of biological interactions in soils and their relationship with functions, as has been done in aquatic environments.

The trophic network and functional trait approaches developed in ecology and applicable to soil communities are considered particularly promising for shifting the focus from the traditional characterisation of ecosystem structure to its functioning. Based on simple morphological criteria (e.g. the shape and visibility of mouthparts, or the colour and size of organisms), these approaches have led to the formation of ecological or functional groups comprising different taxa (e.g. nematodes, springtails, mites and earthworms). Trait-based approaches also overcome the constraints associated with species-level identification, which is often challenging. They also facilitate research on microbes by using the characterisation of soil DNA, which often has difficulty linking taxonomic affiliation to functions or groups of functions. However, these approaches remain limited to describing the living components of ecosystems, and to our knowledge, they have never been applied to soil on the basis of its functions by integrating its biotic and abiotic components.

75. Edaphon: all organisms living in the soil.

Agreeing on a metric for overall pollutant load

Pollutant concentrations are rarely included in overall soil quality indices, instead constituting a separate set of indicators. The most widely studied pollutants are Trace Metals and Metalloids (TMMs), as well as certain organic pollutants such as PAHs and pesticides. So-called emerging pollutants, such as micro- and nanoplastics, antibiotics, manufactured nanoparticles and PFAS, are rarely, if ever, considered. Some comprehensive pollution approaches rely on biological indicators using organisms, such as snails with the Sum of Excess Transfer indicator or plants with the Total Metal Load indicator. These organisms are sensitive enough to assess contamination by pollutants and their bioavailability. Therefore, there is still scope for research into defining indicators that can be used to estimate the overall pollutant load in situations of diffuse pollution.

Developing non-destructive techniques

Proximal detection and remote sensing allow frequent measurements to be taken on the same soil unit without destruction, at spatial resolutions that sampling alone cannot achieve. However, both methods remain dependent on the direct measurement of soil properties in order to calibrate the models used to interpret the information contained in the detected signal. In the case of remote sensing, measurements only concern the surface horizon.

Nevertheless, the sensitivity of proximal detection and remote sensing signals to numerous soil parameters makes them an interesting choice of covariate in soil monitoring and digital mapping approaches. Both techniques are particularly relevant for the initial identification of situations requiring further investigation, based on samples. Another avenue to explore is interpreting the signal in ways that do not involve deconvolution to obtain basic properties, but instead analyse it directly as an indicator of soil health.

Defining the good ecological status of soil

As in other fields dealing with the concept of health, particularly human health, "good health" remains difficult to define. In fact, it is more common to identify a point at which deterioration occurs (i.e. when a disease is diagnosed) than to define what constitutes "good health". For instance, soils in heath vegetation formations are acidic and nutrient-poor, yet they are not considered to be in poor ecological condition. These soils play an important role in carbon storage, and altering their characteristics through fertilisation and liming could lead to the loss of these rare vegetation formations. In order to facilitate the definition of good ecological status, the current literature on soils lacks conceptual reflection on the thresholds corresponding to a deterioration in functions, as well as on the target corresponding to this good ecological status.

Given the diversity of spatial and temporal scales at which the various processes contributing to a single function occur and the constraints to which the soil is subjected, a dual theoretical and operational approach is required for measuring nested processes.

The frequency of monitoring should be determined by knowledge of the system's operating dynamics. For example, repeated measurements should be taken over time in consistently similar soil and climate contexts, focusing on the inter-seasons (spring and autumn) when biological activity is most intense. However, how can this frequency be determined in advance, and do these optimal measurement frequencies actually exist? Are they compatible with socio-economic imperatives?

Assessing the benefits and limitations of proxies

In the face of the current unresolved problems of arbitrating between different functions or aggregating all functions, a pragmatic approach would be to select an indicator to represent optimal soil functioning. The literature reveals that carbon is sometimes already adopted for this purpose. However, although carbon is an important determinant of the biological component in soils (abundance, biomass and diversity), it does not take into account the taxonomic richness that directly supports the diversity of soil functions (i.e. richness, which could also serve as a proxy). Therefore, multifunctionality would be considered indirectly when ensuring that an increase in the carbon storage function does not compromise other functions. Certain biological indices have also been proposed to reflect the overall level of soil functionality (e.g. Soil Biological Quality [QBS-ar], based on soil microarthropods). However, such indices are limited because they exclude microorganisms, nematodes and earthworms, thus not taking into account the entire range of biodiversity. Furthermore, to consider biodiversity as a proxy for functions, it is necessary to document the relationship between the two more thoroughly and deploy research programmes to improve functional ecology knowledge.

Territorialising soil quality monitoring

Work on soil quality governance emphasises the importance of territorialising public action at a local or regional level to ensure consistent implementation of measures and stakeholder involvement. Additionally, the proposed European Soil Monitoring and Resilience Directive, initially published in July 2023, establishes the delineation of soil districts for governance purposes and soil units for monitoring purposes within each Member State. However, the definition of these entities is still very open at this stage. Scientific literature dealing with the delimitation of relevant geographical areas and associated duties highlights different approaches based on: existing administrative units (e.g. SCOTs), which allow consistency with territorial planning but do not necessarily correspond to soil quality realities; units that are homogeneous in terms of soil type and use (small agricultural regions [PRA] are the closest example); units relevant for integrated environmental considerations, for which watersheds appear to be good candidates.

Assessing the legal levers for an integrated approach to soil quality

The intended territorial objectives also call for further research into the extent to which existing planning documents can incorporate soil health goals. In this regard, are the measures resulting from the implementation of the Climate and Resilience Law sufficient, or are they limited? Furthermore, could this not be an opportunity to improve the right to environmental information on soil, as well as citizen participation in soil governance? Currently, the right to information in this area is largely still limited to the register of polluted soils and urban planning. Although this right is rooted in constitutionally protected principles enshrined in the Environmental Charter, it has thus far been overlooked. More consideration needs to be given to the establishment of democratic bodies and services dedicated to this area of governance, based on data systems that ensure citizens are informed and actions are effective.

Would the resulting regulation of land use constitute an unjustified infringement of private property rights? In light of society's growing recognition of the importance of preserving soil quality, it would be interesting to see how the legal obstacles that have previously been identified as hindering any attempt to change the law evolve or even fade away. In reality, these obstacles are more political in nature.

Building upon participatory mechanisms

Participatory science and research programmes improve the consideration of knowledge due to their inclusive approach to developing, selecting and using indicators. They also enable fine-grained data collection in various contexts, including private spaces that are difficult for researchers and institutions to access. However, their main weakness is the challenge of controlling the sampling and measurement methods. However, this issue is partly offset when programmes are widely deployed, as the large volume of data collected enables statistical methods to be applied to identify anomalies.

In this regard, the involvement of French stakeholders in these participatory schemes, and their links with the European Network of Living Labs (ENOLL), could be strengthened with regard to soils. Such initiatives are most successful when a diverse range of stakeholders are involved, including land managers, technology providers, service providers, relevant institutional actors and professional or residential end users. It is also important to have a clear division of roles between scientists, users, and policymakers. This type of system has the advantage of producing results in real conditions or contexts on issues shared by all stakeholders, while being subject to monitoring and evaluation in terms of knowledge gains and ecological, social, and economic benefits.

Appendices

Table A1. Main recent institutional initiatives aimed at establishing a set of soil quality indicators at different governance levels 142

Table A2. Existing values for selected indicators 146

Table A3. Measurement methods considered to analyse the operationality of indicators 155

Table A4. Modalities to quantify the operational criteria of indicators 158

Table A1. Main recent institutional initiatives aimed at establishing a set of soil quality indicators at different governance levels.

	Period	Project (*: with development of a decision-support tool). Reference documents or websites
World	2006-present	**FAO Soils Portal** https://www.fao.org/soils-portal/soil-degradation-restoration/global-soil-health-indicators-and-assessment/global-soil-health/en/
	2015-present	**FAO – Status of the World's Soil Resources – GLADIS** (Global Land Degradation Information System) https://www.fao.org/documents/card/en/c/c6814873-efc3-41db-b7d3-2081a10ede50/
European Union	2023-2026	**Al4SoilHealth*** https://cordis.europa.eu/project/id/101086179
	2023present	**JRC – EUSO** (EU soil observatory) dashboard https://esdac.jrc.ec.europa.eu/esdacviewer/euso-dashboard/
	2022	**EEA** – Soil monitoring in Europe https://www.eea.europa.eu/publications/soil-monitoring-in-europe
	2023	**BENCHMARKS** – European network for the characterisation and harmonisation of monitoring approaches https://soilhealthbenchmarks.eu/
	2021	**EJP Soil – MINOTAUR** – Modelling and mapping soil biodiversity patterns and functions across Europe https://ejpsoil.eu/soil-research/minotaur
	2021-2024	**EJP Soil – SERENA** – Soil ecosystem services and soil threats modelling and mapping https://ejpsoil.eu/soil-research/serena
	2021-2022	**EJP Soil – SIREN** – Stocktaking for agricultural soil quality and ecosystem services indicators and their reference values Faber *et al.*, 2022
	2020	**INSENSE** (Indicators of sensitivity of forest ecosystems subject to increased biomass harvesting) FOR-EVAL - Forest soil assessment* Augusto *et al.*, 2018 https://librairie.ademe.fr/dechets-economie-circulaire/1262-insense-indicateurs-de-sensibilite-des-ecosystemes-forestiers-soumis-a-une-recolte-accrue-de-biomasse.html https://www6.bordeaux-aquitaine.inrae.fr/ispa/Outils/Outils-d-aide-a-la-decision/ For-Eval-a mobile application for assessing forest soils
	2015-2020	**iSQAPER** (interactive soil quality assessment)* https://www.isqaper-is.eu/
	2015-2019	**LANDMARK** consortium* https://landmark2020.eu/list-of-deliverables/
	2011-2014	**ECOFINDERS** (Ecological function and biodiversity indicators in European soils) https://projects.au.dk/ecofinders
	2006-2008	**ENVASSO** (Environmental Assessment of Soil for Monitoring) https://esdac.jrc.ec.europa.eu/Projects/Envasso/documents/ENV_Vol-I_Final2_web.pdf

Highest resolution	Main target users	Types of quality approaches	Types of land use	Types of parameters measured
1 km	public policies	health	all	physical, chemical
< 10 m	public policies	quality, health, threats, safety, functions, ecosystem services, fertility	all	physical, chemical, biological
< 10 m	public policies	health, threats	all	physical, chemical, biological
10 m	public policies	threats	all	physical, chemical, biological
10 m	public policies	quality, health, threats, functions, ecosystem services, fertility	all	physical, chemical, biological
depending on case studies	public policies, farmers	health	all	physical, chemical, biological
10 m	public policies	functions, ecosystem services	agricultural	biological
10 m	public policies	threats, ecosystem services	agricultural	physical, chemical, biological
10 m	public policies	quality, health, threats, safety, functions, ecosystem services, fertility	agricultural	physical, chemical, biological
plot	forest	threats, fertility	forestry	physical, chemical
plot	farmers, public policies	threats, functions,	agricultural	physical, chemical, biological
plot	public policies, farmers	functions, ecosystem services	agricultural	physical, chemical, biological
plot	public policies	threats, functions, ecosystem services	all	biological, economic
5 m	public policies	threats	all	physical, chemical, biological

Table A1. (continued)

	Period	Project (*: with development of a decision-support tool). Reference documents or websites
France	2019-present	**Observatoire national de l'artificialisation (National Observatory on Land Artificialisation)** https://artificialisation.developpement-durable.gouv.fr/
	2017-present	**MUSE** – Prendre en compte la multifonctionnalité des sols dans l'aménagement (Integrating soil multifunctionality into urban planning documents) Branchu *et al.*, 2022 https://www.cerema.fr/fr/actualites/prendre-compte-multifonctionnalite-sols-amenagement?folder=4232
	2017	**SUPRA** – Sols Urbains et Projets d'Aménagement (Urban soils and development projects)* Consalès *et al.*, 2022
	2016-2021	**SOILSERV** – Evaluation multi-échelle des services écosystémiques des sols au sein d'agroécosystèmes (Multi-scale assessment of soil ecosystem services within agroecosystems) https://anr.hal.science/search/index/?q=%2A&rows=30&anrProjectReference_s=ANR-16-CE32-0005&anrProjectAcronym_s=Soilserv
	2019	**ADEME** – Méthodologies d'évaluation des fonctions et des services écosystémiques rendus par les sols (Methodologies for assessing ecosystem functions and services provided by soils) https://librairie.ademe.fr/sols-pollues/491-methodologies-d-evaluation-des-fonctions-et-des-services-ecosystemiques-rendus-par-les-sols.html
	2019	**ADEME** – Diagnostic de la qualité des sols agricoles et forestiers (Diagnosis of agricultural and forest soil quality) https://librairie.ademe.fr/produire-autrement/290-diagnostic-de-la-qualite-des-sols-agricoles-et-forestiers.html
	2019	**ONB** – Observatoire national de la biodiversité (National Biodiversity Observatory) https://naturefrance.fr/indicateurs
	2018	**MNHN** – Atlas français des bactéries du sol (French Atlas of Soil Bacteria) https://sciencepress.mnhn.fr/sites/default/files/documents/fr/fiche-pub-hc41.pdf
sub-national	2022	**Mieux aménager avec les sols vivants (Better spatial development and planning with living soils).** Tours Métropole http://sols-vivants.atu37.org/le-referentiel/
	2019	**ARTISOLS.** Occitanie https://www.agro-bordeaux.fr/wp-content/uploads/2021/05/6_IGCS_2021_CNRS_E_Rabot_et_al_ARTISOLS_project_multifunctionality_index_soils.pdf
	2004	**Bioindicateurs de Qualité des Sols - Bioindicators of soil quality** Bispo *et al.*, 2017 (in particular Chapter 9 " Mise en place d'outils et de bio-indicateurs pertinents de la qualité des sols " - Implementation of relevant soil quality tools and bioindicators)
	2014	**DESTISOL*** https://librairie.ademe.fr/sols-pollues/3923-destisol-les-sols-une-opportunite-pour-un- amenagement-urbain-durable.html
	2008	**UQUALISOL ZU** - Préconisation d'utilisation des sols et qualité des sols en zone urbaine et péri-urbaine. Application du bassin minier de Provence – Recommendations for land use and soil quality in urban and peri-urban areas. Provence mining basin. https://librairie.ademe.fr/urbanisme-et-batiment/563-uqualisol-zu-preconisation-d-utilisation-des-sols-et-qualite-des-sols-en-zone-urbaine-et-peri-urbaine.html

Highest resolution	Main target users	Types of quality approaches	Types of land use	Types of parameters measured
plot	public policies, local authorities	artificialisation	urban	land use
plot	local authorities	functions	urban	physical, chemical, biological
plot	developers	functions	urban	physical, chemical, biological
plot	farmers, local authorities	ecosystem services	agricultural	physical, chemical, biological, economic
plot	public policies	functions, ecosystem services	all	physical, chemical, biological
plot	public policies, local authorities, economic actors	quality, functions, ecosystem services	agricultural, forestry	physical, chemical, biological
			agricultural	biological
16 km	public policies, local authorities, farmers	quality	all	organic
plot	local authorities	functions, ecosystem services	agricultural, urban	physical, chemical, biological
1:125,000	local authorities	functions, ecosystem services	all	physical, chemical
plot	public policies	quality, functions	agricultural, forestry	biological
project site	local authorities	functions, ecosystem services	urban	physical, chemical, land use
up to 10 m	local authorities	quality, functions	urban	physical, chemical, biological

Table A2. Existing values for selected indicators.

Indicator	Range of values by land use					Reference
	Crop rotations	Permanent grassland	Woodland	Orchards and perennial shrub crops	All uses	
Depth	Highly variable – depending on the geo-pedoclimatic context					
Erosion rate	Not analysed in this study					
Granulometry – Texture	Highly variable – depending on the geo-pedoclimatic context					
Bulk density (gcm³) **0-30 cm layer**	n = 878 *Min.*: 0.79 *Max.*: 1.96 *Med.*: **1.36** *Mean*: **1.36**	n = 521 *Min.*: 0.44 *Max.*: 1.98 *Med.*: **1.30** *Mean*: **1.29**	n = 672 *Min.*: 0.48 *Max.*: 2.05 *Med.*: **1.24** *Mean*: **1.24**	n = 59 *Min.*: 1.00 *Max.*: 2.04 *Med.*: **1.54** *Mean*: **1.54**	n = 2,144 *Min.*: 0.37 *Max.*: 2.05 *Med.*: **1.32** *Mean*: **1.31**	RMQS,1st campaign Saby *et al.*, 2019; Munera-Echeverri *et al.*, 2024
Bulk density (g/cm³) **30-50 cm layer**	n = 878 *Min.*: 0.87 *Max.*: 2.16 *Med.*: **1.48** *Mean*: **1.48**	n = 521 *Min.*: 0.20 *Max.*: 2.26 *Med.*: **1.44** *Mean*: **1.43**	n = 672 *Min.*: 0.51 *Max.*: 2.34 *Med.*: **1.38** *Mean*: **1.39**	n = 59 *Min.*: 1.06 *Max.*: 2.10 *Med.*: **1.53** *Mean*: **1.55**	n = 2,144 *Min.*: 0.20 *Max.*: 2.34 *Med.*: **1.47** *Mean*: **1.46**	
Roc fragments content (%)	Highly variable – dependent on geo-pedoclimatic context					
Structural stability (MWD) (mm)	n = 102 *Min.*: 0.2 *Max.*: 2.2 *Med.*: **0.6** *Mean.*: **0.7**	n = 36 *Min.*: 0.6 *Max.*: 3.2 *Med.*: **1.9** *Mean*: **1.8**	n = 30 *Min.*: 0.5 *Max.*: 3.3 *Med.*: **2.4** *Mean*: **2.2**	n = 6 *Min.*: 0.5 *Max.*: 1.4 *Med.*: **0.4** *Mean*: **0.5**		RMQS,1st campaign (174 sites) Rabot *et al.*, 2018
Electrical conductivity (σ) (dS/m)	No national data available					
Saturated hydraulic conductivity (Ks) (m/s) **surface horizon**					n = 86 *Min.*: 4.09 × 10^-7 *Max.*: 1.08 × 10^-4 *Med.*: **3.57 × 10^-5** *Geo. mean*: **2.47 × 10^-5**	SOLHYDRO database (data acquired by INRAE / UR Sols, unpublished)

***Air Capacity* (%)**	Not analysed in this study										
Available water capacity (mm/cm) **surface horizon**	n = 29	*Min.*: 0.73 *Max*: 2.20 *Med.*: **1.28** *Mean*: **1.32**	n = 30	*Min.*: 0.47 *Max.*: 2.77 *Med.*: **1.68** *Mean*: **1.71**	n = 29	*Min.*: 0.32 *Max.*: 5.09 *Med.*: **1.27** *Mean*: **1.48**	n = 3	*Min.*: 0.67 *Max.*: 1.87			RMQS, 2[e] campaign RU-RMQS programme (2016-2023 data)
Available water capacity (mm/cm) **deep horizons**	n = 31	*Min.*: 0.36 *Max.*: 2.47 *Med.*: **1.05** *Mean*: **1.13**	n = 24	*Min.*: 0.62 *Max.*: 3.03 *Med.*: **1.18** *Mean*: **1.38**	n = 37	*Min.*: 0.22 *Max.*: 2.46 *Med.*: **1.04** *Mean*: **1.13**	n = 2	*Min.*: 0.65 *Max.*: 0.95			
***Water Holding Capacity* (WHC)**	Not analysed in this study										
pH (water) layer 0-30 cm	n = 878	*Min.*: 4.5 *Max.*: 8.6 *Med.*: **7.1**	n = 521	*Min.*: 4.5 *Max.*: 8.6 *Med.*: **5.9**	n = 672	*Min*: 3.7 *Max*: 8.7 *Med.*: **5.0**	n = 59	*Min*: 4.9 *Max*: 8.7 *Med.*: **8.1**	n = 2,144	*Min.*: 3.7 *Max.*: 9.2 *Med.*: **6.2**	RMQS, 1[st]campaign Saby *et al.*, 2019; Munera- Echeverri *et al.*, 2024
pH (water) 30-50 cm layer	n = 791	*Min.*: 4.7 *Max.*: 9.1 *Med.*: **7.3**	n = 420	*Min.*: 4.2 *Max.*: 8.9 *Med.*: **6.2**	n = 501	*Min.*: 4.0 *Max.*: 9.0 *Med.*: **5.0**	n = 50	*Min.*: 5.0 *Max.*: 8.8 *Med.*: **8.1**	n = 1,775	*Min.*: 4.0 *Max.*: 9.1 *Med.*: **6.5**	
CEC (meq/100 g soil) **0-30 cm layer**	n = 878	*Min.*: 0.5 *Max.*: 59.6 *Med.*: **11.8** *Mean*: **14.0**	n = 521	*Min.*: 0.8 *Max.*: 63.5 *Med.*: **10.0** *Mean*: **14.5**	n = 672	*Min.*: 0.3 *Max.*: 70.1 *Med.*: **6.7** *Average*: **13.6**	n = 59	*Min.*: 3.6 *Max.*: 39.3 *Med.*: **13.1** *Mean*: **14.6**	n = 2,144	*Min.*: 0.3 *Max.*: 70.1 *Med.*: **10.2** *Mean*: **14.0**	RMQS, 1[st]campaign Saby *et al.*, 2019; Munera- Echeverri *et al.*, 2024
CEC (meq/100 g soil) **30-50 cm layer**	n = 791	*Min.*: 0.8 *Max.*: 62.6 *Med.*: **11.4** *Mean*: **13.2**	n = 420	*Min.*: 1.1 *Max.*: 56.0 *Med.*: **8.3** *Mean*: **11.7**	n = 501	*Min.*: 0.3 *Max.*: 58.3 *Med.*: **5.0** *Mean*: **9.0**	n = 59	*Min.*: 3.4 *Max.*: 34.7 *Med.*: **12.5** *Mean*: **13.9**	n = 1,775	*Min.*: 0.3 *Max.*: 62.2 *Med.*: **8.9** *Mean*: **11.6**	
Total N content (mg/kg)	n = 877	*Min.*: 0.28 *Max*: 5.7 *Med.*: **1.46** *Mean*: **1.65**	n = 521	*Min.*: 0.52 *Max.*: 12.3 *Med.*: **2.43** *Mean*: **2.8**	n = 672	*Min.*: 0.03 *Max*: 11.9 *Med.*: **1.85** *Mean*: **2.33**	n = 59	*Min.*: 0.37 *Max.*: 4.02 *Med.*: **0.9** *Mean*: **1.11**	n = 2,144	*Min.*: 0.03 *Max.*: 16 *Med.*: **1.73** *Mean*: 2.14	RMQS, 1[st]campaign Saby *et al.*, 2019; Munera- Echeverri *et al.*, 2024

Table A2. (continued)

Indicator	Range of values by land use										Reference
	Crop rotations		Permanent grassland		Woodland		Orchards and perennial shrub crops		All uses		
Available P content (Olsen P) (mg/kg) **0-30 cm layer**	n = 878	*Min.:* 0.005 *Max:* 0.557 *Med.:* **0.075** *Mean:* **0.084**	n = 521	*Min.:* 0.001 *Max.:* 1.110 *Med.:* **0.031** *Mean:* **0.047**	n = 672	*Min.:* 0.001 *Max.:* 0.210 *Med.:* **0.009** *Mean:* **0.015**	n = 59	*Min.:* 0.003 *Max.:* 0.162 *Med.:* **0.036** *Mean:* **0.055**	n = 2144	*Min.:* 0.001 *Max.:* 1.110 *Med.:* **0.034** *Mean:* **0.053**	RMQS, 1st campaign Saby *et al.*, 2019; Munera- Echeverri *et al.*, 2024
Available P content (Olsen P) (mg/kg) **30-50 cm layer**	n = 878	*Min.:* 0.001 *Max:* 0.308 *Med.:* **0.022** *Mean:* **0.033**	n = 521	*Min.:* 0.000 *Max.:* 1.060 *Med.:* **0.011** *Mean:* **0.022**	n = 672	*Min.:* 0.001 *Max:* 0.259 *Med.:* **0.005** *Mean:* **0.010**	n = 59	*Min.:* 0.003 *Max.:* 0.134 *Med.:* **0.017** *Mean:* **0.030**	n = 2144	*Min.:* 0.000 *Max.:* 1.060 *Med.:* **0.012** *Mean:* **0.024**	
Available K content (mg/kg)	n = 878	*Min.:* 0.06 *Max:* 2.15 *Med.:* **0.4** *Mean:* **0.46**	n = 521	*Min.:* 0.04 *Max.:* 9.77 *Med.:* **0.30** *Mean:* **0.39**	n = 672	*Min.:* 0.13 *Max.:* 1.08 *Med.:* **0.19** *Mean:* **0.24**	n = 59	*Min.:* 0.11 *Max.:* 1.87 *Med.:* **0.47** *Mean:* **0.52**	n = 2144	*Min.:* 0.01 *Max.:* 9.77 *Med.:* **0.316** *Mean:* **0.376**	RMQS, 1st campaign Saby *et al.*, 2019; Munera-Echeverri *et al.*, 2024
Total TMMs content (mg/kg)	Available in the study report for As, Cd, Co, Cr, Cu, Hg, Ni, Pb, Ti, Zn										
Partial (extractable) content in TMMs (mg/kg)	Specific data are measured as part of the QUASAPROVE programme run by the RMT Al-Chimie										
Organic pollutants content: PAHs (µg/kg)	n = 889	*Min.:* 6.05 *Max.:* 31,193 *Med.:* **90.1** *Mean:* **287**	n = 532	*Min.:* 5.1 *Max.:* 6,521 *Med.:* **73.8** *Mean:* **245**	n = 582	*Min.:* 5.2 *Max.:* 2,656 *Med.:* **60.6** *Mean:* **129**	n = 66	*Min.:* 10.2 *Max.:* 1313 *Med.:* **138** *Mean:* **278**	n = 2154	*Min.:* 5.1 *Max.:* 31,193 *Med.:* **76.3** *Mean:* **229**	RMQS, 1st campaign
Organic pollutant content: PCB (µg/kg)	n = 887	*Min.:* 0.02 *Max.:* 2,867 *Med.:* **0.68** *Mean:* **6.5**	n = 532	*Min.:* 0.02 *Max.:* 17,404 *Med.:* **0.53** *Mean:* **36**	n = 578	*Min.:* 0.02 *Max.:* 63.5 *Med.:* **0.81** *Average:* **1.40**	n = 66	*Min.:* 0.02 *Max:* 54.3 *Med.:* **0.62** *Mean:* **2.01**	n = 2152	*Min.:* 0.02 *Max.:* 17,404 *Med.:* **0.67** *Mean:* **12.0**	

Organic pollutant content: dioxins, furans (ng/kg)	n = 887	*Min.*: 1.02 *Max*: 148 *Med.*: **7.06** *Mean*: **17.6**	n = 532	*Min.*: 1.03 *Max.*: 297 *Med.*: **7.06** *Mean*: **16.1**	n = 579	*Min.*: 1.52 *Max.*: 277 *Med.*: **5.96** *Mean*: **12.0**	n = 66	*Min.*: 1.25 *Max*: 143 *Med.*: **4.38** *Mean*: **14.0**	n = 2153	*Min.*: 1.02 *Max.*: 297 *Med.*: **6.37** *Mean*: **15.4**	RMQS, 1st campaign
Organic pollutant content: pesticides (fungicides) (µg/kg)	n = 29	*Min.*: 0.04 *Max*: 263.7 *Med.*: **24.7** *Mean*: **44.1**	n = 5	*Min.*: 0.08 *Max.*: 5.4 *Med.*: **0.4** *Mean*: **1.56**	n = 5	*Min.*: 0.07 *Max.*: 0.94 *Med.*: **0.21** *Mean*: **0.34**	n = 4	*Min.*: 0.04 *Max.*: 229.8 *Med.*: **1.04** *Mean*: **58.0**	n = 47	*Min.*: 0.04 *Max.*: 264 *Med.*: **7.49** *Mean*: **33.9**	
Organic pollutant content: pesticides (herbicides) (µg/kg)	n = 29	*Min.*: 8.06 *Max*: 1,167.3 *Med.*: **61.1** *Mean*: **128.9**	n = 5	*Min.*: 4.42 *Max.*: 75.2 *Med.*: **33.1** *Mean*: **36.5**	n = 5	*Min.*: 0.05 *Max*: 606.6 *Med.*: **0.95** *Mean*: **202.5**	n = 4	*Min.*: 13.2 *Max.*: 283.5 *Med.*: **24.8** *Mean*: **107.2**	n = 47	*Min.*: 0.05 *Max.*: 1167 *Med.*: **45.7** *Average*: **116**	RMQS, 2e Phytosol Project campaign (47 sites)
Organic pollutant content: pesticides (insecticides) (µg/kg)	n = 29	*Min.*: 0.21 *Max*: 136.7 *Med.*: **1.41** *Mean*: **10.1**	n = 5	*Min.*: - *Max.*: - *Med.*: - *Mean*: -	n = 5	*Min.*: 0.26 *Max.*: 0.26 *Med.*: **0.26** *Mean*: **0.26**	n = 4	*Min.*: 0.82 *Max.*: 4.73 *Med.*: **1.09** *Mean*: **2.21**	n = 47	*Min.*: 0.21 *Max.*: 413 *Med.*: **23.3** *Mean*: **1.38**	
Organic carbon content (g/kg) **0-30 cm layer**	n = 878	*Min.*: 2.6 *Max*: 58.2 *Med.*: **14.6** *Mean*: **16.7**	n = 521	*Min.*: 6.8 *Max.*: 145 *Med.*: **25.2** *Mean*: **29.4**	n = 672	*Min.*: 0.6 *Max*: 170 *Med.*: **28.2** *Mean*: **34.9**	n = 59	*Min.*: 3.4 *Max.*: 39.3 *Med.*: **9.2** *Mean*: **11.6**	n = 2,144	*Min.*: 0.6 *Max.*: 243 *Med.*: **20.4** *Mean*: **25.6**	RMQS, 1st campaign Saby *et al.*, 2019; Munera- Echeverri *et al.*, 2024
Organic carbon content (g/kg) **30-50 cm layer**	n = 792	*Min.*: 0.8 *Max*: 52 *Med.*: **6.5** *Mean*: **8.1**	n = 420	*Min.*: 1.8 *Max*: 142 *Med.*: **8.7** *Mean*: **11.8**	n = 501	*Min.*: 0.6 *Max.*: 128 *Med.*: **8.5** *Mean*: **13.1**	n = 50	*Min.*: 2.0 *Max.*: 17 *Med.*: **5.7** *Mean*: **6.9**	n = 1,775	*Min.*: 0.6 *Max.*: 235 *Med.*: **7.4** *Mean*: **10.6**	
Corg/Clay ratio	n = 877	*Min.*: 0.012 *Max*: 0.54 *Med.*: **0.03** *Mean*: **0.08**	n = 521	*Min.*: 0.03 *Max.*: 0.72 *Med.*: **0.11** *Mean*: **0.13**	n = 672	*Min.*: 0.01 *Max.*: 9 *Med.*: **0.14** *Mean*: **0.23**	n = 59	*Min.*: 0.01 *Max.*: 0.16 *Med.*: **0.04** *Mean*: **0.05**	n = 2,144	*Min.*: 0.01 *Max.*: 9 *Med.*: **0.09** *Mean*: **0.14**	RMQS,1st campaign Saby *et al.*, 2019; Munera-Echeverri *et al.*, 2024

Table A2. (continued)

Indicator	Range of values by land use										Reference
	Crop rotations		Permanent grassland		Woodland		Orchards and perennial shrub crops		All uses		
Carbon stock (kg C/m²) **0-30 cm layer**	n = 884	*Min.*: 1.0 *Max.*: 13.7 *Med.*: **4.8** *Mean*: **5.2**	n = 601	*Min.*: 1.8 *Max.*: 30.9 *Med.*: **7.8** *Mean*: **8.5**	n = 586	*Min.*: 0.7 *Max.*: 23.0 *Med.*: **7.3** *Mean*: **8.1**	n = orchard: 22; vineyard: 42	*Min.*: orchard: 1.6 vineyard: 0.5 *Max*: orchard: 9.8 vineyard: 6.3 *Med*: orchard: **4.5** vineyard: **3.2** *Mean*: orchard: **4.7** vineyard: **3.4**	*Mean*: **6.85** (weighted average based on the number of entries in each category)		RMQS, 1st campaign
Carbon stock (kg C/m²) **0-50 cm layer**		Data currently being consolidated							n = 2,144	*Min.*: 0.5 *Max.*: 40.7 *Med.*: **7.6** *Mean*: **8.6**	RMQS, 1st campaign
Carbon fractions: physical coarse, POC (g C/kg soil)	n = 405	*Min.*: 0.72 *Max.*: 13.85 *Med.*: **1.96** *Mean*: **2.37**	n = 246	*Min.*: 1.13 *Max.*: 55.85 *Med.*: **4.36** *Mean*: **6.04**	n = 243	*Min.*: 1.13 *Max.*: 36.97 *Med.*: **6.69** *Mean*: **9.09**	n = 26	*Min.*: 0.86 *Max.*: 10.12 *Med.*: **1.84** *Mean*: **2.68**	n = 960	*Min.*: 0.37 *Max.*: 55.85 *Med.*: **3.35** *Mean*: **5.38**	Delahaie *et al.*, 2024
Carbon fractions: physical fine, MAOC (g C/kg soil)	n = 405	*Min.*: 3.25 *Max*: 46.27 *Med.*: **12.56** *Mean*: **14.20**	n = 246	*Min.*: 6.81 *Max.*: 69.83 *Med.*: **20.94** *Mean*: **23.89**	n = 243	*Min.*: 3.16 *Max.*: 82.08 *Med.*: **18.21** *Mean*: **21.76**	n = 26	*Min.*: 4.02 *Max*: 35.97 *Med.*: **7.11** *Mean*: **10.23**	n = 960	*Min.*: 0.22 *Max.*: 82.08 *Med.*: **15.84** *Mean*: **18.87**	

Carbon fractions: thermal active, C_a (g C/kg soil)	n = 785	*Min.:* 1.70 *Max.:* 32.53 *Med.:* **7.23** *Mean:* **8.52**	n = 479	*Min.:* 2.75 *Max.:* 84.86 *Med.:* **15.31** *Mean:* **18.74**	n = 520	*Min.:* 0.31 *Max.:* 83.14 *Average:* **£16.95** *Mean:* **21.39**	n = 42	*Min.:* 0.70 *Max:* 18.93 *Med.:* **3.16** *Mean:* **4.40**	n = 1,879	*Min.:* 0.31 *Max.:* 84.86 *Med.:* **11.56** *Mean:* **15.04**	Delahaie *et al.*, 2024
Carbon fractions: thermal stable, C_s (g C/kg soil)	n = 785	*Min.:* 2.71 *Max.:* 24.35 *Med.:* **7.53** *Mean:* **8.52**	n = 479	*Min.:* 3.25 *Max.:* 44.81 *Med.:* **9.65** *Mean:* **11.45**	n = 520	*Min.:* 1.17 *Max.:* 38.13 *Med.:* **9.52** *Mean:* **11.95**	n = 42	*Min.:* 2.71 *Max.:* 12.50 *Med.:* **6.59** *Mean:* **7.07**	n = 1,879	*Min.:* 1.17 *Max.:* 44.89 *Med.:* **8.43** *Mean:* **10.25**	
Oxidisable carbon fraction	No national data available										
C/N ratio 0-10 cm layer					n = 102	*Min.:* 12.8 *Max:* 40.3 *Mean:* **20.1**					RENECOFOR data in Achat *et al.*, 2018
C/N ratio 0-30 cm layer	n = 877	*Min.:* 5.1 *Max.:* 22.6 *Med.:* **9.9** *Mean:* **10.1**	n = 521	*Min.:* 8.1 *Max.:* 17.4 *Med.:* **10.3** *Mean:* **10.5**	n = 672	*Min.:* 7.5 *Max.:* 52.7 *Med.:* **15.1** *Mean:* **16.5**	n = 59	*Min.:* 5.2 *Max.:* 19.0 *Med.:* **10.2** *Mean:* **10.6**	n = 2,144	*Min.:* 5.1 *Max.:* 52.7 *Med.:* **10.8** *Mean:* **12.3**	RMQS, 1st campaign Saby *et al.*, 2019 and 2024
C/N ratio 30-50 cm layer	n = 791	*Min.:* 3.0 *Max:* 26.5 *Med.:* **9.1** *Mean:* **8.9**	n = 420	*Min.:* 4.9 *Max.:* 30.2 *Med.:* **9.2** *Mean:* **9.6**	n = 501	*Min.:* 4.0 *Max.:* 57.0 *Med.:* **13.2** *Mean:* **14.8**	n = 50	*Min.:* 3.5 *Max.:* 19.0 *Med.:* **9.4** *Mean:* **9.8**	n = 1,775	*Min.:* 3.0 *Max.:* 57.0 *Med.:* **9.5** *Mean:* **10.9**	
Type of forest humus					n = 102	Mull: 50%. Moder/Amphi: 40% *Mor*: 10%					RENECOFOR data and Zanella *et al.*, 2011
Microbial molecular biomass DNA (µg/g soil)	n = 878	*Min.:* 0.4 *Max.:* 306.2 *Mean:* **38.5**	n = 521	*Min.:* 2.3 *Max.:* 455 *Mean:* **84.7**	n = 672	*Min.:* 0.2 *Max.:* 629.6 *Mean:* **76.8**	n = 59	*Min.:* 0.2 *Max.:* 249.2 *Mean:* **24.4**	n = 2,144	*Min.:* 1.1 *Max:* 412.5 *Mean:* **75.3**	RMQS, 1st campaign

Table A2. (continued)

Indicator	Range of values by land use										Reference
	Crop rotations		Permanent grassland		Woodland		Orchards and perennial shrub crops		All uses		
Microbial biomass (mg C/kg soil)	n = 885	*Min.*: 0.2 *Max.*: 306.2 *Mean:* **38.5**	n = 536	*Min.*: 2.3 *Max.*: 455 *Mean:* **84.7**	n = 584	*Min.*: 0.2 *Max.*: 629.6 *Mean:* **76.4**	n = 42	*Min.*: 0.2 *Max.*: 249.2 *Mean:* **23.6**	n = 2,146	*Min.*: 0.2 *Max.*: 629.6 *Mean:* **61.7**	RMQS,1[st]campaign Karimi *et al.*, 2018 *Atlas français des bactéries du sol* Horrigue *et al.*, 2016
Basal soil respiration (Oxitrop)	n = 83	*Min.*: 0 *Max:* 1.312 *Mean:* **0.486**	n = 24	*Min.*: 0.667 *Max:* 1.493 *Mean:* **1.023**	n = 20	*Min.*: 0.183 *Max.*: 1.251 *Mean:* **0.612**			n = 187	*Min.*: 0 *Max.*: 1.633 *Mean:* **0.638**	Programme Bio-indicateurs – Phase 2
N and C mineralisation potential	Currently being acquired as part of the RMQS biodiversity programme										
Total PLFA (nmol/g dry soil)	n = 84	*Min.*: 35.9 *Max.*: 534.6 *Mean:* **146.9**	n = 24	*Min.*: 160.6 *Max.*: 1,846.6 *Mean:* **527.4**	n = 20	*Min.*: 68.6 *Max:* 813.23 *Mean:* **268.66**			n = 188	*Min.*: 23.8 *Max.*: 1,846.6 *Mean:* **247.2**	Programme Bio-indicateurs – Phase 2
Soil bacteria density (millions of 16S rDNA/g soil)	n = 842	*Min.*: 163 *Max:* 499,978 *Mean:* **92,627**	n = 463	*Min.*: 473 *Max.*: 50,193 *Mean:* **14,148**	n = 490	*Min.*: 136 *Max.*: 50,564 *Mean:* **10,152**	n = 47	*Min.*: 200 *Max.*: 22,737 *Mean:* **6,185**	n = 1,842	*Min.*: 136 *Max.*: 50,600 *Mean:* **10,648**	RMQS, 1[st]campaign Djemiel *et al.*, 2023
Soil bacterial diversity (number of operational taxonomic units)	n = 743	*Min.*: 1,192 *Max:* 3,075 *Mean:* **2,181**	n = 470	*Min.*: 1,141 *Max.*: 2,901 *Mean:* **2,109**	n = 501	*Min.*: 870 *Max.*: 2,962 *Mean:* **1,902**	n = 27	*Min.*: 1,748 *Max.*: 2,579 *Mean:* **2,212**	n = 1,842	*Min.*: 870 *Max.*: 3,075 *Mean:* **2,083**	RMQS,1[st]campaign Karimi *et al.*, 2018 *Atlas français des bactéries du sol* Terrat *et al.*, 2020
Soil fungus density (millions of 18S rDNA/g soil)	n = 847	*Min.*: 3 *Max:* 1,485 *Mean:* **252**	n = 489	*Min.*: 3.9 *Max.*: 1,484 *Mean:* **353.6**	n = 503	*Min.*: 3.1 *Max.*: 527.3 *Mean:* **387.3**	n = 48	*Min.*: 4.1 *Max.*: 1,199.9 *Mean:* **180.4**	n = 1,887	*Min.*: 3.1 *Max.*: 1,527.3 *Mean:* **312.5**	RMQS,1[st]campaign Djemiel *et al.*, 2023 and 2024

Soil fungal diversity (number of operational taxonomic units)	n = 848,848	*Min.:* 703 *Max.:* 2,363 *Mean:* **1,497**	n = 510	*Min.:* 501 *Max:* 2,316 *Mean:* **1,464**	n = 542	*Min.:* 452 *Max.:* 2,577 *Mean:* **1,392**	n = 57	*Min.:* 813 *Max.:* 2,112 *Mean:* **1,370**	n = 2,060	*Min.:* 452 *Max.:* 2,577 *Mean:* **1,454**	RMQS, 1st campaign Djemiel *et al.*, 2024
Fungus/bacteria ratio	n = 849	*Min.:* 0.24 *Max.:* 11.59 *Mean:* **3.05**	n = 509	*Min.:* 0.25 *Max:* 11.75 *Mean:* **2.52**	n = 521	*Min.:* 0.36 *Max.:* 12.15 *Mean:* **4.45**	n = 50	*Min.:* 1.08 *Max.:* 10.75 *Mean:* **3.07**	n = 1,929	*Min.:* 0.24 *Max.:* 12.15 *Mean:* **3.29**	RMQS, 1st campaign Djemiel *et al.*, 2023
Enchytraeids abundance	No national data available										
Earthworms abundance (number of individuals/m²)	n = 262	*Min.:* 0 *Max.:* **917** *Mean:* **223**	n = 117	*Min.:* 32 *Max:* 1333 *Mean:* **421**	n = 20	*Min.:* 90 *Max.:* 219 *Mean:* **92**	n = 126	*Min.:* 0 *Max.:* 1,092 *Mean:* **163**			Forests: Programme Bioindicateurs. Other land uses: OPVT data
Earthworms diversity (number of taxa)	n = 229	*Min.:* 0 *Max.:* 11 *Mean:* **5**	n = 90	*Min.:* 5 *Max.:* 13 *Mean:* **9**	n = 20	*Min.:* 0 *Max.:* 9 *Mean:* **3**	n = 120	*Min.:* 0 *Max.:* 10 *Mean:* **3**			
Total nematodes abundance (number of individuals/g dry soil)	n = 3,367	*Min.:* 0.05 *Max.:* 113 *Mean:* **16.6**	n = 1,215	*Min.:* 0 *Max.:* 149 *Mean:* **18.1**	n = 483	*Min.:* 0.24 *Max.:* 132 *Mean:* **14.0**	n = 1,111	*Min.:* 0.06 *Max.:* 256 *Mean:* **11.4**	n = 6,176	*Min.:* 0 *Max.:* 256 *Mean:* **15.8**	ELIPTO database (ELISOL Environnement Database)
Total nematode diversity (Shannon index)	n = 3,367	*Min.:* 0.19 *Max:* 2.69 *Mean:* **1.91**	n = 1,215	*Min.:* 0 *Max.:* 2.78 *Mean:* **1.81**	n = 483	*Min.:* 0.48 *Max.:* 2.84 *Mean:* **1.86**	n = 1,111	*Min.:* 0.21 *Max.:* 2.63 *Mean:* **1.98**	n = 6,176	*Min.:* 0 *Max:* 2.84 *Mean:* **1.90**	
Microarthropods abundance (springtails) (number of individuals/m²)	n = 304	*Min.:* 0 *Max.:* 10.10^3 *Mean:* **4.10^3**	n = 46	*Min.:* 0 *Max.:* 34.10^3 *Mean:* **12.10^3**	n = 25	*Min.:* 0 *Max.:* 27.10^3 *Mean:* **10.10^3**	n = 136	*Min.:* 0 *Max.:* 15,103 *Mean:* **6.10^3**			Meta-analysis Joimel *et al.*, 2017
Micro-arthropods diversity (springtails) (number of species)					n = 25	*Min.:* 0 *Mean:* **8.6**	n = 136	*Min.:* 0 *Mean:* **3.2**			

Table A2. (continued)

Indicator	Range of values by land use									Reference
	Crop rotations		Permanent grassland		Woodland		Orchards and perennial shrub crops	All uses		
Microarthropods Abundance (mites) (number of individuals/m²)	n = 47	*Min.*: 707 *Max.*: 16,154 *Mean:* **4,756**	n = 42	*Min.*: 825 *Max.*: 21,460 *Mean:* **7,835**	n = 8	*Min.*: 5,188 *Max.*: 39,500 *Mean:* **12,823**		n = 98	*Min.*: 707 *Max.*: 39,500 *Mean:* **6,755**	RMQS – Biodiv 1 – Brittany
Microarthropods diversity (mites)	No national data available									
Enzymatic activities	Currently being acquired at certain RMQS sites (200 points)									
Enzymatic activities Galactosidase (mU/g dry soil8)	n = 84	*Min.*: 0 *Max:* 18.23 *Mean:* **4.57**	n = 24	*Min.*: 1.86 *Max.*: 7.88 *Mean:* **3.57**	n = 20	*Min.*: 2.41 *Max.*: 15.00 *Mean:* **6.19**		n = 188	*Min.*: 0 *Max.*: 39.25 *Mean:* **5.56**	
Enzymatic activities β- Glucosidase (nmol/min/g dry soil)	n = 84	*Min.*: 1.17 *Max.*: 23.88 *Mean:* **7.63**	n = 24	*Min.*: 2.84 *Max.*: 19.00 *Mean:* **9.73**	n = 20	*Min.*: 0 *Max.*: 10.52 *Mean:* **3.57**		n = 188	*Min.*: 0 *Max.*: 25.64 *Mean:* **7.79**	
Enzymatic activities Arylsulfatase (mU/g dry soil)	n = 84	*Min.*: 0.65 *Max:* 10.89 *Mean:* **3.50**	n = 24	*Min.*: 2.58 *Max.*: 32.09 *Mean:* **8.6**	n = 20	*Min.*: 0.51 *Max.*: 56.80 *Mean:* **26.81**		n = 188	*Min.*: 0.15 *Max.*: 56.8 *Mean:* **8.86**	Programme Bio-indicateurs – Phase 2
Enzymatic activities Acid phosphatase (nmol/min/g dry soil)	n = 84	*Min.*: 0 *Max:* 56.08 *Mean:* **10.74**	n = 24	*Min.*: 0 *Max.*: 44.19 *Mean:* **19.09**	n = 20	*Min.*: 0 *Max.*: 104.34 *Mean:* **20.44**		n = 188	*Min.*: 0 *Max.*: 104.34 *Mean:* **14.74**	
Enzyme activities Alkaline phosphatase (nmol/min/g dry soil)	n = 84	*Min.*: 2.35 *Max.*: 29.06 *Mean:* **13.23**	n = 24	*Min.*: 2.18 *Max.*: 24.41 *Mean:* **8.53**	n = 20	*Min.*: 0.23 *Max.*: 38.48 *Mean:* **16.87**		n = 188	*Min.*: 0 *Max.*: 41.49 *Mean:* **13.43**	
Ants abundance	Not analysed in this study									
Ants diversity	Not analysed in this study									

Table A3. Measurement methods considered to analyse the operationality of indicators.

Indicator	Basic metric	References	Type of measurement
Organic carbon (content, g/kg)	Organic carbon (content, g/kg)	Several standardised methods	Chemical characterisation
pH in water	pH in water	NF EN ISO 10390	
Available phosphorus (content, g/kg)	Available P (content, g/kg)	Neyroud et Lischer (2003)	
Total N (content, g/kg)	Total N (content, g/kg)	ISO 13878 Ros *et al.* (2011)	
CEC	CEC	NF X 31-130	
Inorganic pollutants and metalloids (TMMs)	Total concentrations	Méthodes listées par le GEMAS et l'ICP Forests : Attaque acide NF X31-147 Acide fluorhydrique – acide perchlorique NF ISO 14869-1 ; SA12A Fusion alcaline ISO 14869-2 ; SA12B	
	Pseudo-total concentrations	Eau régale (NF ISO 11466/NF EN ISO 54321 ; GLOSOLAN-SOP-19 ; SA11) suivi des méthodes analytiques SAA, ICP-OES (NF EN ISO 11885), ICP-MS (EN ISO 17294-2). Méthode spécifique Hg : ISO 16772- AAS vapeur froide	
	Partial concentrations and agricultural context	Extraction au DTPA : NF X 31-121 ; GLOSOLAN-SOP	
	Partial concentrations and transfer	Diverses extractions Cipullo *et al.* (2018)	
	Partial concentrations and health accessibility	Bioaccessibilité orale : NF ISO 17924 ; UBM du groupe européen BARGE Par ex., Li *et al.* (2021) ; Billmann *et al.* (2023)	
Available K (content, g/kg)	Available K (content, g/kg)	Afnor (1999). Recueil de normes, Qualité des sols, vol. 1, p. 566.	

Table A3. (continued)

Indicator	Basic metric	References	Type of measurement
Labile carbon/fraction	Physical fractionation	NF X 31-516	Chemical characterisation
	Thermal fractionation	Delahaie *et al.* (2024)	
	Permanganate-oxidisable fraction	Weil *et al.* (2009)	
Electrical conductivity	Ece	GLOSOLAN-SOP-08	
	Ec 1/5	ISO 11265 ; GLOSOLAN-SOP-07	
Organic pollutants (content, µg/kg)	Pesticides and metabolites	Froger *et al.* (2023)	
	PCBs - dioxins - furans		
	PAHs	NF ISO 18287	
Structural stability	Weighted average diameter (Le Bissonnais method)	NF X31-515/ISO 10930	Physical
Particle size distribution	Particle size	NF X31-107	
Depth (m)	Depth (m)		
Humus	Humus	Zanella *et al.* (2018)	
Bulk density (kg/L)	Bulk density (kg/L)		
Saturated hydraulic conductivity	Saturated hydraulic conductivity		
Available water capacity (mm)	Measured	ISO 11274	
Available water capacity (mm)	Estimated by pedotransfer functions	Dobarco *et al.* (2019) Roman Dobarco *et al.* (2019)	Estimated
Cstock (t carbon)	0-30 cm	Méthode FAO (2019c) Intègre teneur en carbone, densité apparente et profondeur	

C/N ratio	C/N ratio		**Esti-mated**
Corg/clay ratio	Corg/clay ratio	Plusieurs méthodes normées (EEA, 2023)	
Microbiology (abundance and diversity)	Microbial molecular biomass (µg DNA/gsol)	NF ISO 11063, intégrant les modifications de Terrat *et al.* (2017) Horrigue *et al.* (2016)	**Biological characterisation**
	Microbial balance (fungi/bacteria ratio)	NF ISO 11063 Djemiel *et al.* (2024)	
	Bacterial diversity (number of taxa)	NF ISO 11063 Terrat *et al.*, 2017	
	Fungal diversity (number of taxa)	NF ISO 11063 Djemiel *et al.* (2024)	
	Amount of microbial carbon and nitrogen (microbial biomass)	NF ISO 14240-2	
Nematodes (abundance and diversity)	Total abundance and functional groups	ISO 23611-4	
	Nematode indices	ISO 23611-4	
Microarthropods and mesofauna (abundance and diversity)	Abundance	ISO 23611-2	
	Abundance of functional groups	ISO 23611-2	
	Taxonomic richness and diversity	ISO 23611-2	
Enchytraedes (abundance)	Abundance	ISO 23611-3	
Lombrichids (abundance and diversity)	Abundance	ISO 23611-1	
	Abundance of functional groups	ISO 23611-1	
	Taxonomic richness and diversity	ISO 23611-1	
Enzymatic activity	Dehydrogenase, beta-glucosidase, phosphatases, arylsulfatase, ureases	NF EN ISO 20130 Trevors (1984) Méthode colorimétrique ou fluorogène ISO 22939	**Biological activity characterisation**
Basal soil respiration	Basal soil respiration	Norme ISO 16072	
N and C mineralisation potential	N and C mineralisation potential	NF XP U44-163	

Table A4. Modalities to quantify the operational criteria of indicators.

Group	Criterion	Type	Level	Score
Technical operationality – Credibility	Scientific and conceptual basis	Binary	No	0
			Yes	1
	Valid and standardised to describe the state/functions of the ecosystem	Hierarchical	0- Not measurable or no standardised technique	0
			1- Measurable by a standardised research-level technique with reliable measurement precise and robust (*Tier* 3)	1
			2- Measurable using a standardised laboratory technique with reliable, accurate and robust measurement (*Tier* 2)	2
Technical operationality – Feasibility	Technical feasibility	Binary	No	0
			Yes	1
	Existence of databases	Hierarchical	0- No data available	0
			1- Availability of data with limited time and/or spatial coverage (*Tier* 2)	1
			2- Data available nationwide and/or covering long periods of time (*Tier* 1)	2
Technical operationality – Salience	Sensitive and deployable for monitoring	Hierarchical	0- No	0
			1- Yes, but significant constraints to implementation (sampling) given high intra-plot or intra-annual variability	1
			2- Yes, but mainly for spatial monitoring (low temporal variability)	2
			3- Yes for spatial and temporal monitoring	3
	Comprehensible and usable for carrying out an assessment/ diagnosis in order to facilitate decision-making	Hierarchical	0- No existing values, thresholds or targets that can be used to establish a benchmark and/or are not comprehensible	0
			1- Understandable and associated with a reference framework for interpretation restricted to certain territories / time periods (*Tier* 2 level)	1
			2- Understandable and associated with a national interpretation framework covering long periods of time (*Tier* 1 level)	2

Table A4. (continued)

Group	Criterion	Type	Level	Score
Use-related operationality – Salience	Relevance of information and user awareness for action	Prioritised	0- No	0
			1- Yes, but appropriation by non-scientific users requiring intermediation (scientists, advisers, etc.) for the interpretation and implementation of actions	1
			2- Yes, with at least partial appropriation by non-scientific users (farmers, elected officials, etc.)	2
	Use of indicators to meet a specific objective	Binary	No	0
			Yes	1
Use-related operationality – Legitimacy	Acceptance of indicators	Binary	No	0
			Yes	1
Use-related operationality – Transferability (framework) and modularity (scale)	Implementation framework Individual initiative / research / participatory research or action research / service provision	Combination of situations	No situation	0
			1 situation	1
			2 situations	2
			3 situations	3
			4 situations	4
	Steering of actions resulting from the use of indicators Local level and/or regional level	Combination of situations	None	0
			Local	1
			Territorial	1
			Local and regional	2

Bibliographic sources cited

The complete list of references analysed in the study report can be consulted online: https://doi.org/10.17180/qnpx-x742.

AbdelRahman M.A.E., 2023. An overview of land degradation, desertification and sustainable land management using GIS and remote sensing applications. *Rendiconti Lincei-Scienze Fisiche E Naturali*, 34 (3), 767-808, https://doi.org/10.1007/s12210-023-01155-3.

Achat D.L., Pousse N., Nicolas M., Augusto L., 2018. Nutrient remobilization in tree foliage as affected by soil nutrients and leaf life span. *Ecological Monographs*, 88 (3), 408-428, https://doi.org/10.1002/ecm.1300.

Adhikari K., Smith D.R., Collins H., Hajda C., Acharya B.S., Owens P.R., 2022. Mapping within-field soil health variations using apparent electrical conductivity, topography, and machine learning. *Agronomy-Basel*, 12 (5), 16, https://doi.org/10.3390/agronomy12051019.

Andrews S.S., Karlen D.L., Cambardella C.A., 2004. The soil management assessment framework: A quantitative soil quality evaluation method. *Soil Science Society of America Journal*, 68 (6), 1945-1962, https://doi.org/10.2136/sssaj2004.1945.

Anthony M.A., Bender S.F., Van der Heijden M.G.A., 2023. Enumerating soil biodiversity. *Proceedings of the National Academy of Sciences of the United States of America*, 120 (33), 9, https://doi.org/10.1073/pnas.2304663120.

Antoni V., Arrouays D., Bispo A., Brossard M., Le Bas C., Stengel P. *et al.*, 2011. *L'État des sols de France*, Paris, groupement d'intérêt scientifique sur les Sols, https://www.gissol.fr/publications/rapport-sur-letat-des-sols-de-france-2-849.

Aronson J., Floret C., Le Floc'h E., Ovalle C., Pontanier R., 1993. Restoration and rehabilitation of degraded ecosystems in arid and semi-arid lands. I. A View from the South. *Restoration Ecology*, 1 (1), 8-17, https://onlinelibrary.wiley.com/doi/abs/10.1111/j.1526-100X.1993.tb00004.x.

Arrouays D., Stengel P., Feix I., Lesaffre B., Morard V., Bardy M. *et al.*, 2022. Le GIS Sol, sa genèse et son évolution au cours des vingt dernières années. *Étude et gestion des sols*, 29, 365-379, https://www.afes.fr/ressources/le-gis-sol-sa-genese-et-son-evolution-au-cours-des-vingt-dernieres-annees.

Augusto L., Pousse N., Achat D., Brédoire F., Bronner T., Durante S. *et al.*, 2018. Projet INSENSÉ : « Indicateurs de sensibilité des écosystèmes forestiers soumis à une récolte accrue de biomasse ». *Comment préserver les sols forestiers dans un contexte de récolte accrue de bois ?* Hal INRAE, https://hal.inrae.fr/hal-03193840.

Bai Z., Li H., Yang X., Zhou B., Shi X., Wang B. *et al.*, 2013. The critical soil P levels for crop yield, soil fertility and environmental safety in different soil types. *Plant Soil*, 372, 27-37, https://doi.org/10.1007/s11104-013-1696-y.

Baize D., Girard M.C., 2009. *Référentiel pédologique 2008*, Versailles, Quæ, 405 p, https://edepot.wur.nl/481543.

Baize D., Jabiol B., 2012. *Guide pour la description des sols*. Nouvelle édition. Éditions Quae. 448 p.

Balestrat M., Barbe É., Chery J.-P., Lagacherie P., Tonneau J.-P., 2011. Reconnaissance du patrimoine agronomique des sols : une démarche novatrice en Languedoc-Roussillon. *Norois. Environnement, aménagement, société* (221), 83-96, https://doi.org/10.4000/norois.3752.

Ball B.C., Batey T., Munkholm L.J., 2007. Field assessment of soil structural quality – a development of the Peerlkamp test. *Soil Use and Management*, 23 (4), 329-337, https://doi.org/10.1111/j.1475-2743.2007.00102.x.

Bampa F., O'Sullivan L., Madena K., Sandén T., Spiegel H., Henriksen C.B. *et al.*, 2019. Harvesting European knowledge on soil functions and land management using multi-criteria decision analysis. *Soil Use and Management*, 35 (1), 6-20, https://doi.org/10.1111/sum.12506.

Bartkowski B., Bartke S., 2018. Leverage points for governing agricultural soils: A review of empirical studies of European farmers' decision-making. *Sustainability*, 10 (9), 3179, https://doi.org/10.3390/su10093179.

Bartkowski B., Droste N., Ließ M., Sidemo-Holm W., Weller U., Brady M.V., 2021. Payments by modelled results: A novel design for agri-environmental schemes. *Land Use Policy*, 102, 105230, https://doi.org/10.1016/j.landusepol.2020.105230.

Baveye P.C., Baveye J., Gowdy J., 2016. Soil "Ecosystem" services and natural capital: Critical appraisal of research on uncertain ground. *Frontiers in Environmental Science*, 4 (41), https://doi.org/10.3389/fenvs.2016.00041.

Beck H.E., Zimmermann N.E., McVicar T.R., Vergopolan N., Berg A., Wood E.F., 2018. Present and future Köppen-Geiger climate classification maps at 1-km resolution. *Scientific Data*, 5 (1), 180214, https://doi.org/10.1038/sdata.2018.214.

Bhaduri D., Sihi D., Bhowmik A., Verma B.C., Munda S., Dari B., 2022. A review on effective soil health bio-indicators for ecosystem restoration and sustainability. *Frontiers in Microbiology*, 13, 25, https://doi.org/10.3389/fmicb.2022.938481.

Billet P., 2016a. La prise en compte de la qualité des sols dans le droit français, in : Bispo A., Guellier C., Martin E., Sapijanskas J., Soubelet H., Chenu C., eds. Les Sols. *Intégrer leur multifonctionnalité pour une gestion durable*, Versailles, Quæ, 259 p. (coll. Savoir faire).

Billet P., 2016b. *NORMASOL. Recherches sur la protection juridique des fonctions et services du sol. Programme Gessol (MEDDE/Ademe)* : ministère de l'Écologie et du développement durable, Ademe, https://univ-lyon3.hal.science/hal-02126279.

Billmann M., Hulot C., Pauget B., Badreddine R., Papin A., Pelfrêne A., 2023. Oral bioaccessibility of PTEs in soils: A review of data, influencing factors and application in human health risk assessment. *Science of the Total Environment*, 896, 20, https://doi.org/10.1016/j.scitotenv.2023.165263.

Bispo A., Jolivet C., Ranjard L., Cluzeau D., Hedde M., Peres G., 2017. Mise en place d'outils et bio-indicateurs pertinents de qualité des sols. *Les Sols et la vie souterraine : des enjeux majeurs en agroécologie*, Versailles, Quæ, 328 p., https://hal.science/hal-01606253.

Bispo A., Schnebelen N., 2018. Synthèse des outils, indicateurs, référentiels disponibles pour comprendre et piloter la biologie des sols. *Innovations agronomiques*, 69, 91-100, https://doi.org/10.15454/DCZJMP.

Blaikie P., 2016. *The Political Economy of Soil Erosion in Developing Countries*, Londres, Routledge, 200 p., https://doi.org/10.4324/9781315637556.

Blanchart A., Séré G., Cherel J., Warot G., Stas M., Consalès J.N. *et al.*, 2018. Towards an operational methodology to optimize ecosystem services provided by urban soils. *Landscape and Urban Planning*, 176, 1-9, https://doi.org/10.1016/j.landurbplan.2018.03.019.

Boizard H., Peigné J., Vian J.-F., Duparque A., Tomis V., Johannes A. *et al.*, 2019. Visual methods for assessing soil structure in support of a clinical approach in agronomy. *Agronomie, environnement & sociétés*, 9 (2), 55-76, https://isara.hal.science/hal-04103319.

Bongers T., 1990. The maturity index – an ecological measure of environmental disturbance based in nematode species composition. *Oecologia*, 83 (1), 14-19, https://doi.org/10.1007/bf00324627.

Bonilla-Bedoya S., Valencia K., Herrera M.A., López-Ulloa M., Donoso D.A., Pezzopane J.E.M., 2023. Mapping 50 years of contribution to the development of soil quality biological indicators. *Ecological Indicators*, 148, 13, https://doi.org/10.1016/j.ecolind.2023.110091.

Bouleau G., Deuffic P., 2016. Qu'y a-t-il de politique dans les indicateurs écologiques ? *VertigO*, 16 (2), https://doi.org/10.4000/vertigo.17581.

Boutet D., Serrano J., 2013. Les sols périurbains, diversification des activités et des valeurs. Quelques éléments de comparaison et d'analyse. Économie rurale. *Agricultures, alimentations, territoires*, (338), 5-23, https://doi.org/10.4000/economierurale.4142.

Branchu P., Marseille F., Béchet B., Bessière J.P., Boithias L., Duvigneau C. *et al.*, 2022. MUSE. *Intégrer la multifonctionnalité dans les documents d'urbanisme*, rapport et annexes, 219 p., https://librairie.ademe.fr/urbanisme-et-batiment/5415-muse-integrer-la-multifonctionnalite-des-sols-dans-les-documents-d-urbanisme.html.

Brauch H.G., Spring Ú.O., 2009. Securizing the ground, grounding security, Desertification land degradation and drought. *UNCCD issue paper*, 2, 52, http://library.unccd.int/Details/fullCatalogue/843.

Brauman A., Thoumazeau A., 2020. Biofunctool® : un outil de terrain pour évaluer la santé des sols, basé sur la mesure de fonctions issues de l'activité des organismes du sol. *Étude et gestion des sols*, 27 (1), 289-304, https://www.afes.fr/ressources/biofunctool-un-outil-de-terrain-pour-evaluer-la-sante-des-sols-base-sur-la-mesure-de-fonctions-issues-de-lactivite-des-organismes-du-sol.

Brunet Y., Voltz M., 2023. Les sols : éléments d'un cycle dynamique, in : Euzen A., Eymard L., Gaill F. (eds), *Le Développement durable à découvert*, Paris, CNRS Éditions, 78-79, https://doi.org/10.4000/books.editionscnrs.10626.

Buisson E., 2011. *Community and restoration ecology, importance of disturbance, natural resilience and assembly rules*, thèse d'habilitation à diriger les recherches, Université d'Avignon et des Pays de Vaucluse.

Bünemann E.K., Bongiorno G., Bai Z.G., Creamer R.E., De Deyn G., de Goede R. *et al.*, 2018. Soil quality — A critical review. *Soil Biology & Biochemistry*, 120, 105-125, https://doi.org/10.1016/j.soilbio.2018.01.030.

Cipullo S., Prpich G., Campo P., Coulon F., 2018. Assessing bioavailability of complex chemical mixtures in contaminated soils: Progress made and research needs. *Science of the Total Environment*, 615, 708-723, https://doi.org/10.1016/j.scitotenv.2017.09.321.

Clarimont B., Barbe É., Lagacherie P., 2021. Enjeux autour des terres agricoles et des données pédologiques : point de vue opérationnel d'un service de l'État en région, in : Brennan J., Maurel P., Plant R. (eds), *Les Terres agricoles face à l'urbanisation : De la donnée à l'action, quels rôles pour l'information ?*, chapitre 2, Versailles, Quæ, 51-64 (coll. Update Sciences & Technologie), http://books.openedition.org/quae/28335.

Compagnone C., Pribetich J., 2017. Quand l'abandon du labour interroge les manières d'être agriculteur : Changement de norme et diversité des modèles d'agriculture. *Revue française de socio-économie*, 18 (1), 101-121, https://doi.org/10.3917/rfse.018.0101.

Consalès J.-N., Blanchart A., Séré G., Vidal-Beaudet L., Schwartz C., 2022. Le sol, une ressource à considérer dans les stratégies d'aménagement des villes : mise en place d'une démarche collaborative pour construire un outil d'aide à la décision d'affectation des sols. *Revue scientifique sur la conception et l'aménagement de l'espace*, 27, https://doi.org/10.4000/paysage.31354.

Cousin I., Desrousseaux M., Leenhardt S., coord., 2024. *Préserver la qualité des sols : vers un référentiel d'indicateurs*. Synthèse du rapport d'étude, INRAE (France), 130 p. https://doi.org/10.17180/qnpx-x742.

Czúcz B., Keith H., Maes J., Driver A., Jackson B., Nicholson E., Kiss M., Obst C., 2021. Selection criteria for ecosystem condition indicators. *Ecological Indicators*, 133, 108376, https://doi.org/10.1016/j.ecolind.2021.108376.

de Bello F., Lavorel S., Hallett L.M., Valencia E., Garnier E., Roscher C. *et al.*, 2021. Functional trait effects on ecosystem stability: Assembling the jigsaw puzzle. *Trends in Ecology & Evolution*, 36 (9), 822-836, https://doi.org/10.1016/j.tree.2021.05.001.

De Gruijter J.J., Brus D.J., Bierkens M.F.P., Knotters M., 2006. *Sampling for Natural Resource Monitoring*, Heidelberg, Springer.

Delahaie A.A., Cécillon L., Stojanova M., Abiven S., Arbelet P., Arrouays D. *et al.*, 2024. Investigating the complementarity of thermal and physical soil organic carbon fractions. *EGUsphere*, 2024, 1-25, https://doi.org/10.5194/egusphere-2024-197.

Djemiel C., Dequiedt S., Horrigue W., Bailly A., Lelièvre M., Tripied J. *et al.*, 2024. Unraveling biogeographical patterns and environmental drivers of soil fungal diversity at the French national scale. *Soil*, 10 (1), 251-273, https://doi.org/10.5194/soil-10-251-2024.

Dobarco M.R., Cousin I., Le Bas C., Martin M.P., 2019. Pedotransfer functions for predicting available water capacity in French soils, their applicability domain and associated uncertainty. *Geoderma*, 336, 81-95, https://doi.org/10.1016/j.geoderma.2018.08.022.

Dokoutchaiev V.V., 1883. *Tchernoziom russe*, these. Saint-Petersburg. 480 p. https://www.persee.fr/doc/rhs_0151-4105_1983_num_36_3_1942.

Dominati E.J., Mackay A.D., Bouma J., Green S., 2016. An ecosystems approach to quantify soil performance for multiple outcomes: the future of land evaluation? *Soil Science Society of America Journal*, 80 (2), 438-449, https://doi.org/10.2136/sssaj2015.07.0266.

Domínguez-Haydar Y., Velásquez E., Carmona J., Lavelle P., Chavez L.F., Jiménez J.J., 2019. Evaluation of reclamation success in an open-pit coal mine using integrated soil physical, chemical and biological quality indicators. *Ecological Indicators*, 103, 182-193, https://doi.org/10.1016/j.ecolind.2019.04.015.

Donadieu P., Rémy E., Girard M.-C., 2016. Les sols peuvent-ils devenir des biens communs ? *Natures Sciences Sociétés*, 24 (3), 261-269, https://doi.org/10.1051/nss/2016025.

Donnars C., Tibi A., Caillaud M.A., Dashkina R., Girard A., Leenhardt S. *et al.*, 2021. *Principes de conduite des expertises scientifiques collectives et des études en éclairage des politiques et du débat publics* (version 2, novembre 2021), INRAE DEPE, 63 p., https://doi.org/10.15454/tkna-4w25.

Doran J.W., Zeiss M.R., 2000. Soil health and sustainability: Managing the biotic component of soil quality. *Applied Soil Ecology*, 15 (1), 3-11, https://doi.org/10.1016/s0929-1393(00)00067-6.

Duflo E., 2017. The economist as plumber. *American Economic Review*, 107 (5), 1-26, https://doi.org/10.3386/w23213.

Ehrlich P., Walker B., 1998. Roundtable: Rivets and redundancy. *BioScience*, 48 (5), 387-387, https://doi.org/10.2307/1313377.

EEA (European Environment Agency), 2023. Soil monitoring in Europe — Indicators and thresholds for soil quality assessments. *Publications Office of the European Union*, https://doi.org/10.2800/956606.

Faber J.H., Cousin I., Meurer K.H.E., Hendriks C.M.J., Bispo A., Viketoft M. *et al.*, 2022. *Stocktaking for Agricultural Soil Quality and Ecosystem Services Indicators and their Reference Values*, EJP SOIL Internal Project SIREN Deliverable 2, Report, 153.

Fadil A., El Wahidi F., 2023. Vers une nouvelle classification des modèles d'évaluation et de prédiction de l'érosion hydrique. *Physio-Géo*, 19, https://doi.org/10.4000/physio-geo.15783.

Faith D.P., 2021. Valuation and appreciation of biodiversity: Thẹ "Maintenance of options" provided by the variety of life. *Frontiers in Ecology and Evolution*, 9, https://doi.org/10.3389/fevo.2021.635670.

FAO, 2006. *Guidelines for soil description* — Fourth edition, Rome, Food and Agriculture Organization of the United Nations, 109 p., https://www.fao.org/4/a0541e/a0541e.pdf.

FAO, 2019a. *Measuring and modelling soil carbon stocks and stock changes in livestock production systems: Guidelines for assessment* (Version 1) — Livestock Environmental Assessment and Performance (LEAP) Partnership, Rome, FAO, 170 p., https://www.fao.org/3/CA2934EN/ca2934en.pdf.

FAO, 2019b. *Outcome document of the Global Symposium on Soil Erosion*, Rome, FAO, 28 p., https://openknowledge.fao.org/server/api/core/bitstreams/e5717cbe-ac10-414e-bcb9-33aeb87fbbed/content.

FAO, 2019c. *Standard Operating Procedure for Soil Organic Carbon Walkley-Black Method*: Rome. Titration and Colorimetric Method. https://www.fao.org/3/ca7471en/ca7471en.pdf.

Legros, 1996. *Cartographie des Sols. De l'analyse spatiale à la gestion des territoires.* EPFL Press, Lausanne. 370 p. https://www.epflpress.org/produit/418/9782880742980/cartographies-des-sols.

Farinetti A., 2013. La protection juridique de la qualité du sol au prisme du droit de l'eau. *Environnement et développement durable*, n° 6, étude 17, 143-157.

Fine A.K., Van Es H.M., Schindelbeck R.R., 2017. Statistics, scoring functions, and regional analysis of a comprehensive soil health database. *Soil Science Society of America Journal*, 81 (3), 589-601, https://doi.org/10.2136/sssaj2016.09.0286.

Forsyth T., 2007. Sustainable livelihood approaches and soil erosion risks: Who is to judge? *International Journal of Social Economics*, 34 (1/2), 88-102, https://doi.org/10.1108/03068290710723381.

Fournil J., Kon Kam King J., Granjou C., Cécillon L., 2018. Le sol : enquête sur les mécanismes de (non-)émergence d'un problème public environnemental. *VertigO*, 18 (2), https://doi.org/10.4000/vertigo.20433.

Friedlingstein P., O'Sullivan M., Jones M.W., Andrew R.M., Bakker D.C.E., Hauck J. *et al.*, 2023. Global Carbon Budget 2023. *Earth System Science Data*, 15 (12), 5301-5369, https://doi.org/10.5194/essd-15-5301-2023.

Froger C., Jolivet C., Budzinski H., Pierdet M., Caria G., Saby N.P.A. *et al.*, 2023. Pesticide residues in French soils: Occurrence, risks, and persistence. *Environmental Science & Technology*, 57 (20), 7818-7827, https://doi.org/10.1021/acs.est.2c09591.

Gann G.D., McDonald T., Walder B., Aronson J., Nelson C.R., Jonson J. *et al.*, 2019. International principles and standards for the practice of ecological restoration, 2nd edition. *Restoration Ecology*, 27, S3-S46, https://doi.org/10.1111/rec.13035.

Gascuel-Odoux C., Renault P., Antoni V., Arrouays D., Bougon N., Denys S. *et al.*, 2023. Quelles perspectives scientifiques et techniques pour l'inventaire et la surveillance des sols en France : Quels besoins en données, comment mieux les acquérir, les diffuser, les utiliser ? *Étude et gestion des sols*, https://www.afes.fr/wp-content/uploads/2023/04/EGS_2023_30_Gascuel_51-64.pdf.

Gauthier N., 2020. Écologiser une frontière agri-urbaine dans le Val-de-Marne. *Revue d'anthropologie des connaissances*, 14 (4), https://doi.org/10.4000/rac.12616.

Gautronneau Y., Manichon H., 1987. *Guide méthodique du profil cultural*, CEREF-ISARA/GEARA-INAPG, https://agriressources.fr/fileadmin/user_upload/Auvergne-Rhone-Alpes/177_Eve-agriressources/fertisols/RESSOURCES/Diagnostic/ISARA_Lyon__guide_test_de_beche__9_.pdf.

Girard M.-C., Schvartz C., Jabiol B., 2023. Les sols de la terre, in Girard M.-C., Schvartz C., Jabiol B., *Étude des sols*, chapitre 1, Paris, Dunod, 1-12, https://stm.cairn.info/etude-des-sols--9782100857449?lang=fr.

Gouée P., Marie M., Cantat O., Bensaïd A., 2010. DÉMÉTER : une démarche originale pour maîtriser la consommation du foncier agricole liée à l'étalement urbain. *OPDE Outils pour Décider Ensemble*, Inra Montpellier, 25-26 octobre 2010, https://www.researchgate.net/publication/49136410_DEMETER_une_demarche_originale_pour_maitriser_la_consommation_du_foncier_agricole_liee_a_l'etalement_urbain.

Greiner L., Keller A., Grêt-Regamey A., Papritz A., 2017. Soil function assessment: review of methods for quantifying the contributions of soils to ecosystem services. *Land Use Policy*, 69, 224-237, https://doi.org/10.1016/j.landusepol.2017.06.025.

Griffiths B.S., Römbke J., Schmelz R.M., Scheffczyk A., Faber J.H., Bloem J. *et al.*, 2016. Selecting cost effective and policy-relevant biological indicators for European monitoring of soil biodiversity and ecosystem function. *Ecological Indicators*, 69, 213-223, https://doi.org/10.1016/j.ecolind.2016.04.023.

Grimonprez B., 2019. Le droit au service de la justice foncière du xxie siècle, in Dominique P., Pierre B., Benoît G. (eds), *La Terre en commun : plaidoyer pour une justice foncière*, Paris, Fondation Jean-Jaurès, https://hal.science/hal-02916253.

Guimaraes R.M.L., Ball B.C., Tormena C.A., 2011. Improvements in the visual evaluation of soil structure. *Soil Use and Management*, 27 (3), 395-403, https://doi.org/10.1111/j.1475-2743.2011.00354.x.

Haines-Young R., Potschin M., 2010. The links between biodiversity, ecosystem services and human well-being, in Raffaelli D.G., Frid C.L.J. (eds), *Ecosystem Ecology: A new synthesis*, Cambridge, CUP, https://doi.org/10.1017/CBO9780511750458.007.

Hajer M.A., 2005. *The Politics of Environmental Discourse: Ecological modernization and the policy process*, Oxford, Clarendon Press, 344 p.

Hallsworth E.G., 1987. *Anatomy, Physiology and Psychology of Erosion*, Chichester, J. Wiley (IFIAS monograph).

Haslmayr H.P., Geitner C., Sutor G., Knoll A., Baumgarten A., 2016. Soil function evaluation in Austria — Development, concepts and examples. *Geoderma*, 264, 379-387, https://doi.org/10.1016/j.geoderma.2015.09.023.

Hassenteufel P., 2010. Les processus de mise sur agenda : sélection et construction des problèmes publics. *Informations sociales*, 157 (1), 50-58, https://doi.org/10.3917/inso.157.0050.

Heal G.M., Daily G., Ehrlich P., Salzman J.E., Boggs C., Hellman J. *et al.*, 2001. *Protecting natural capital through Ecosystem Service Districts*, SSRN, https://doi.org/10.2139/ssrn.279114.

Hermon C., 2018. La protection du sol en droit, in Hermon C. (éd.), *Services écosystémiques et protection des sols : Analyses juridiques et éclairages agronomiques*, Versailles, Quæ, 326 p., http://books.openedition.org/quae/30685.

Herrick J.E., Jones T.L., 2002. A dynamic cone penetrometer for measuring soil penetration resistance. *Soil Science Society of America Journal*, 66 (4), 1320-1324, https://doi.org/10.2136/sssaj2002.1320.

Hervé-Fournereau N., 2009. Le principe d'intégration. *Droit et politiques de l'environnement*. La Documentation française, 31-40, https://shs.hal.science/halshs-00428707.

Horrigue W., Dequiedt S., Prévost-Bouré N.C., Jolivet C., Saby N.P.A., Arrouays D. *et al.*, 2016. Predictive model of soil molecular microbial biomass. *Ecological Indicators*, 64, 203-211, https://doi.org/10.1016/j.ecolind.2015.12.004.

Houllier F., Joly P.-B., Merilhou-Goudard J.-B., 2017. Dossier Des recherches participatives dans la production des savoirs liés à l'environnement — Les sciences participatives : une dynamique à conforter. *Natures Sciences Sociétés*, 25 (4), 418-423, https://doi.org/10.1051/nss/2018005.

Ineris, 2012. Les polychlorobiphényles - PCB 89.

Inra Unité InfoSol, 2015. *Dictionnaire des données DoneSol 3, Version du 1er novembre 2015*, 470 p., https://dw4.gissol.fr/fichiers/dictionnaire_donesol_igcs_latest.pdf.

Israel-Jost V., 2015. L'attribution d'autorité à la science : Approche néopoppérienne du problème. *Cahiers philosophiques*, 142 (3), 53-72, https://doi.org/10.3917/caph.142.0053.

Jasanoff S., 2010. *States of Knowledge: The co-production of science and social order*, Londres, Routledge, 330 p. (International library of sociology), http://ndl.ethernet.edu.et/bitstream/123456789/17555/1/20.pdf.

Jaunatre R., 2012. *Dynamique et restauration d'une steppe méditerranéenne après changements d'usages (La Crau, Bouches-du-Rhône, France)*, Université d'Avignon, https://theses.hal.science/tel-00862398.

Jeffery S., Verheijen F.G.A., 2020. A new soil health policy paradigm: Pay for practice not performance! *Environmental Science & Policy*, 112, 371-373, https://doi.org/10.1016/j.envsci.2020.07.006.

Joimel S., Schwartz C., Hedde M., Kiyota S., Krogh P.H., Nahmani J. *et al.*, 2017. Urban and industrial land uses have a higher soil biological quality than expected from physicochemical quality. *Science of the Total Environment*, 584, 614-621. https://doi.org/10.1016/j.scitotenv.2017.01.086.

Jónsson J.Ö.G., Davíðsdóttir B., Jónsdóttir E.M., Kristinsdóttir S.M., Ragnarsdóttir K.V., 2016. Soil indicators for sustainable development: A transdisciplinary approach for indicator development using expert stakeholders. *Agriculture, Ecosystems & Environment*, 232, 179-189, https://doi.org/10.1016/j.agee.2016.08.009.

Juerges N., Hagemann N., Bartke S., 2018. A tool to analyse instruments for soil governance: the REEL-framework. *Journal of Environmental Policy and Planning*, 20 (5), 617-631, https://doi.org/10.1080/1523908X.2018.1474731.

Juerges N., Hansjürgens B., 2018. Soil governance in the transition towards a sustainable bioeconomy — A review. *Journal of Cleaner Production*, 170, 1628-1639, https://doi.org/10.1016/j.jclepro.2016.10.143.

Karimi B., Chemidlin Prévost-Bouré N., Dequiedt S., Terrat S., Ranjard L., 2018. *Atlas français des bactéries du sol*, Paris, Muséum national d'Histoire naturelle, Mèze, Biotope (Hors collection, 41), 192 p., https://sciencepress.mnhn.fr/fr/collections/hors-collection/atlas-francais-des-bacteries-du-sol.

Karlen D.L., Mausbach M.J., Doran J.W., Cline R.G., Harris R.F., Schuman G.E., 1997. Soil quality: A concept, definition, and framework for evaluation. *Soil Science Society of America Journal*, 61 (1), 4-10, https://doi.org/10.2136/sssaj1997.03615995006100010001x.

Keller C., Lambert-Habib M.-L., Robert S., Ambrosi J.-P., Rabot É., 2012. Méthodologie pour la prise en compte des sols dans les documents d'urbanisme : application à deux communes du bassin minier de Provence. *Sud-Ouest européen. Revue géographique des Pyrénées et du Sud-Ouest*, (33), 11-24, https://doi.org/10.4000/soe.173.

Keuskamp J.A., Dingemans B.J.J., Lehtinen T., Sarneel J.M., Hefting M.M., 2013. Tea Bag Index: a novel approach to collect uniform decomposition data across ecosystems. *Methods in Ecology and Evolution*, 4 (11), 1070-1075, https://doi.org/10.1111/2041-210x.12097.

Kibblewhite M.G., Ritz K., Swift M.J., 2008. Soil health in agricultural systems. *Philosophical Transactions of the Royal Society B-Biological Sciences*, 363 (1492), 685-701, https://doi.org/10.1098/rstb.2007.2178.

Kopittke P.M., Menzies N.W., 2007. A review of the use of the basic cation saturation ratio and the "ideal" soil. *Soil Science Society of America Journal*, 71 (2), 259-265. https://doi.org/10.2136/sssaj2006.0186.

Langlais A., 2015. L'appréhension juridique de la qualité des sols agricoles par le prisme des services écosystémiques. *Revue de droit rural*, (435), 28-33, https://univ-rennes.hal.science/hal-01207318.

Laroche B., Almeida Falcon J.-L., Boukir H., Cousin I., Chenu J.-P., Girot G. *et al.*, 2024. Le programme Inventaire gestion et conservation des sols IGCS, in *La Connaissance des sols au services des projets de territoire*, OpenIG, INRAE, avril 2024, Montpellier, https://hal.science/hal-04595742.

Larson W.E., Pierce F.J., 1991. Conservation and enhancement of soil quality. *Evaluation for sustainable land management in the developing world*, 2.

Le Bissonnais Y., 1996. Aggregate stability and assessment of soil crustability and erodibility. 1. Theory and methodology. *European Journal of Soil Science*, 47 (4), 425-437. https://doi.org/10.1111/j.1365-2389.1996.tb01843.x.

Lecerf A., 2017. Methods for estimating the effect of litterbag mesh size on decomposition. *Ecological Modelling*, 362, 65-68, https://doi.org/10.1016/j.ecolmodel.2017.08.011.

Lehmann J., Bossio D.A., Kögel-Knabner I., Rillig M.C., 2020. The concept and future prospects of soil health. *Nature Reviews Earth & Environment*, 110), 544-553, https://doi.org/10.1038/s43017-020-0080-8.

Lévy-Bruhl V., 1992. *La Protection de la faune sauvage en droit français*, thèse, Lyon 3.

Li H., Huang G., Meng Q., Ma L., Yuan L., Wang F. *et al.*, 2011. Integrated soil and plant phosphorus management for crop and environment in China. A review. *Plant and Soil*, 349 (1-2), 157-167. https://doi.org/10.1007/s11104-011-0909-5.

Li J., Zhou D.-d., Chen S., Yan H.-B., Yang X.-Q., 2021. Spatial associations between tree regeneration and soil nutrient in secondary Picea forest in Guandi Mountains, Shanxi, China. *Chinese Journal of Applied Ecology*, 32 (7), 2363-2370, https://doi.org/10.13287/j.1001-9332.202107.006.

Lin Z.B., Schneider A., Sterckeman T., Nguyen C., 2016. Ranking of mechanisms governing the phytoavailability of cadmium in agricultural soils using a mechanistic model. *Plant and Soil*, 399 (1-2), 89-107, https://doi.org/10.1007/s11104-015-2663-6.

Lu Q.F., Liu T.T., Wang N.Q., Dou Z.C., Wang K.G., Zuo Y.M., 2020. A review of soil nematodes as biological indicators for the assessment of soil health. *Frontiers of Agricultural Science and Engineering*, 7 (3), 275-281, https://doi.org/10.15302/j-fase-2020327.

Maliszewska-Kordybach B., 1996. Polycyclic aromatic hydrocarbons in agricultural soils in Poland: preliminary proposals for criteria to evaluate the level of soil contamination. *Appl. Geochemistry*, 11, 121-127. https://doi.org/10.1016/0883-2927(95)00076-3.

Manning P., Van der Plas F., Soliveres S., Allan E., Maestre F.T., Mace G. *et al.*, 2018. Redefining ecosystem multifunctionality. *Nat Ecol Evol*, 2 (3), 427-436, https://doi.org/10.1038/s41559-017-0461-7.

Marschner P., 2012. *Marschner's Mineral Nutrition of Higher Plants*, Londres, Elsevier, Academic Press, https://doi.org/10.1016/C2009-0-63043-9.

Marseille F., Boithias L., Lamari S., 2019. *Quelle prise en compte des sols dans les documents d'urbanisme ?* Paris, Cerema, http://www.cerema.fr/fr/actualites/quelle-prise-compte-sols-documents-urbanisme-premier-rapport.

Martin-Scholz A., Mayere A., Barbe E., Valette E., Maurel P., 2013. Quand l'information échappe à ses créateurs. Le cas de l'artificialisation des terres agricoles en Languedoc-Roussillon. *Études de communication*, 40, https://doi.org/10.4000/edc.5199.

Mazzon M., Cavani L., Ciavatta C., Campanelli G., Burgio G., Marzadori C., 2021. Conventional versus organic management: application of simple and complex indexes to assess soil quality. *Agriculture, Ecosystems & Environment*, 322, 107673, https://doi.org/10.1016/j.agee.2021.107673.

McBratney A., Field D.J., Koch A., 2014. The dimensions of soil security. *Geoderma*, 213, 203-213, https://doi.org/10.1016/j.geoderma.2013.08.013.

McConnell K.E., 1983. An economic model of soil conservation. *American Journal of Agricultural Economics*, 65 (1), 83-89, https://doi.org/10.2307/1240340.

Meynier A., 2020. *Réflexions sur les concepts en droit de l'environnement*, LGDJ, thèse, 624 p.

Mouchet M.A., Paracchini M.L., Schulp C.J.E., Stürck J., Verkerk P.J., Verburg P.H., Lavorel S., 2017. Bundles of ecosystem (dis)services and multifunctionality across European landscapes. *Ecological Indicators*, 73, 23-28, https://doi.org/10.1016/j.ecolind.2016.09.026.

Néel C., Boithias L., Duplanil E., Duvigneau C., Le Guern C., Métois R. *et al.*, 2022. *Qualité des sols et urbanisme. Construire une méthodologie adaptée aux besoins des territoires et favoriser son appropriation*, Ademe, 114 p., https://doc.cerema.fr/Default/doc/SYRACUSE/592638/qualite-des-sols-et-urbanisme-construire-une-methodologie-adaptee-aux-besoins-des-territoires-et-fav?_lg=fr-FR.

Neyroud J.-A., Lischer P., 2003. Do different methods used to estimate soil phosphorus availability across Europe give comparable results? *Journal of Plant Nutrition and Soil Science*, 166 (4), 422-431, https://doi.org/10.1002/jpln.200321152.

Niemeijer D., de Groot R.S., 2008. A conceptual framework for selecting environmental indicator sets. *Ecological Indicators*, 8 (1), 14-25, https://doi.org/10.1016/j.ecolind.2006.11.012.

Nortjé G.P., Laker M.C., 2021. Factors that determine the sorption of mineral elements in soils and their impact on soil and water pollution. *Minerals*, 11 (8), 20, https://doi.org/10.3390/min11080821.

O'Sullivan L., Wall D., Creamer R., Bampa F., Schulte R.P.O., 2018. Functional Land Management: Bridging the Think-Do-Gap using a multi-stakeholder science policy interface. *Ambio*, 47 (2), 216-230, https://doi.org/10.1007/s13280-017-0983-x.

Obriot F., Stauffer M., Goubard Y., Cheviron N., Peres G., Eden M. *et al.*, 2016. Multi-criteria indices to evaluate the effects of repeated organic amendment applications on soil and crop quality. *Agriculture Ecosystems & Environment*, 232, 165-178, https://doi.org/10.1016/j.agee.2016.08.004.

Orgiazzi A., Ballabio C., Panagos P., Jones A., Fernández-Ugalde O., 2018. LUCAS Soil, the largest expandable soil dataset for Europe: a review. *European Journal of Soil Science*, 69 (1), 140-153, https://doi.org/10.1111/ejss.12499.

Ost F., 1995. *La Nature hors la loi. L'écologie à l'épreuve du droit*, Paris, La Découverte, 346 p.

Party J.P., Sauter J., Lux M., Muller N., 2014. Classement des sols et classement des terres pour l'aménagement foncier. Méthodes et adaptation en Alsace. *Étude et gestion des sols*, 21, 61-76, https://www.afes.fr/wp-content/uploads/2023/04/EGS_21_1_2105_Party_61_76.pdf.

Pieper S., Frauenstein J., Ginzky H., Glante F., Grimski D., Kotschik P. *et al.*, 2023. *The upcoming European Soil Health Law — chances and challenges for an effective soil protection*, Scientific opinion paper, German Environment Agency, 37 p., https://www.umweltbundesamt.de/publikationen/the-upcoming-european-soil-health-law-chances.

Piron D., Boizard H., Heddadj D., Pérès G., Hallaire V., Cluzeau D., 2017. Indicators of earthworm bioturbation to improve visual assessment of soil structure. *Soil & Tillage Research*, 173, 53-63, https://doi.org/10.1016/j.still.2016.10.013.

Plant R., Maurel P., Ruoso L.-É., Barbe É., Brennan J., 2021. Synthèse. De la donnée à l'intelligence collective sur les terres agricoles périurbaines : quels rôles pour l'information, les savoirs et l'action ? *Les Terres agricoles face à l'urbanisation : De la donnée à l'action, quels rôles pour l'information ?* Versailles, Quæ, 249-268 (Update Sciences & Technologie), http://books.openedition.org/quae/28505.

Pollock L.J., O'Connor L.M.J., Mokany K., Rosauer D.F., Talluto M.V., Thuiller W., 2020. Protecting biodiversity (in all its complexity), new models and methods. *Trends in Ecology & Evolution*, 35 (12), 1119-1128, https://doi.org/10.1016/j.tree.2020.08.015.

Pollock L.J., Thuiller W., Jetz W., 2017. Large conservation gains possible for global biodiversity facets. *Nature*, 546 (7656), 141-144, https://doi.org/10.1038/nature22368.

Prager K., Schuler J., Helming K., Zander P., Ratinger T., Hagedorn K., 2011. Soil degradation, farming practices, institutions and policy responses: An analytical framework. *Land Degradation & Development*, 22 (1), 32-46, https://doi.org/10.1002/ldr.979.

Pülzl H., Rametsteiner E., 2009. Indicator development as 'boundary spanning' between scientists and policy-makers. *Science and Public Policy*, 36 (10), 743-752, https://doi.org/10.3152/030234209X481987.

Rabot E., Guiresse M., Pittatore Y., Angelini M., Keller C., Lagacherie P., 2022. Development and spatialization of a soil potential multifunctionality index for agriculture (Agri-SPMI) at the regional scale. Case study in the Occitanie region (France). *Soil Security*, 6, 100034, https://doi.org/10.1016/j.soisec.2022.100034.

Ranjard L., 2016. *AgrInnov Tester les indicateurs de l'état biologique des sols en lien avec les pratiques agricoles*, https://agriressources.fr/fileadmin/user_upload/Auvergne-Rhone-Alpes/177_Eve-agriressources/fertisols/RESSOURCES/Diagnostic/Nov2015-Comifer-Gemas-ARTICLE-RANJARD.pdf.

Ranjard L., 2020. Sciences participatives au service de la qualité écologique des sols. *Génie écologique*, https://doi.org/10.51257/a-v1-ge1074.

Renzi G., Canfora L., Salvati L., Benedetti A., 2017. Validation of the soil Biological Fertility Index (BFI) using a multidimensional statistical approach: A country-scale exercise. *Catena*, 149, 294-299, https://doi.org/10.1016/j.catena.2016.10.002.

Richards L.A., 1954. Diag*nosis and Improvement of Saline and Alkali Soils*. U.S. Gov. Printing Office, Washington, DC.

Richelle L., 2019. *De la fertilité des sols à la santé de la terre. Retour sur un processus d'apprentissage collectif visant l'évaluation de la santé des sols cultivés en agriculture paysanne*, thèse, Université de Namur, https://pure.unamur.be/ws/portalfiles/portal/41280240/2019_RichelleL_these.pdf.

Richer-De-Forges A.C., Baffet M., Berger C., Coste S., Courbe C., Jalabert S. *et al.*, 2014. La cartographie des sols à moyennes échelles en France métropolitaine. *Étude et Gestion des Sols*, 21, 25-36, https://www.afes.fr/ressources/la-cartographie-des-sols-a-moyennes-echelles-en-francemetropolitaine.

Robert S., Ajmone-Marsan F., Ambrosi J.P., Biasioli M., Cormier C., Criquet S. *et al.*, 2013. *Préconisation d'utilisation des sols et qualité des sols en zone urbaine et péri-urbaine — Application du bassin minier de Provence*, rapport de recherche, Ademe, 273 p., https://hal.science/hal-01787627.

Robinson D.A., Fraser I., Dominati E.J., Davíðsdóttir B., Jónsson J.O.G., Jones L. *et al.*, 2014. On the value of soil resources in the context of natural capital and ecosystem service delivery. *Soil Science Society of America Journal*, 78 (3), 685-700, https://doi.org/10.2136/sssaj2014.01.0017.

Roman Dobarco M., Cousin I., Le Bas C., Martin M., 2019. Pedotransfer functions for predicting available water capacity in French soils, their applicability domain and associated uncertainty. *Geoderma*, 336, 81-95, https://doi.org/10.1016/j.geoderma.2018.08.022.

Ros G.H., Temminghoff E.J.M., Hoffland E., 2011. Nitrogen mineralization: a review and meta-analysis of the predictive value of soil tests. *European Journal of Soil Science*, 62 (1), 162-173, https://doi.org/10.1111/j.1365-2389.2010.01318.x.

Rossiter D.G., 1995. Economic land evaluation: why and how. *Soil Use and Management*, 11 (3), 132-140, https://doi.org/10.1111/j.1475-2743.1995.tb00511.x.

Ruiz F., Cherubin M.R., Ferreira T.O., 2020. Soil quality assessment of constructed Technosols: Towards the validation of a promising strategy for land reclamation, waste management and the recovery of soil functions. *Journal of Environmental Management*, 276, 11, https://doi.org/10.1016/j.jenvman.2020.111344.

Saby N., Bertouy B., Boulonne L., Bispo A., Ratié C., Jolivet C., 2019. *Statistiques sommaires issues du RMQS sur les données agronomiques et en éléments traces des sols français de 0 à 50 cm*. Recherche Data Gouv. https://doi.org/10.15454/BNCXYB.

Schulte R.P.O., Bampa F., Bardy M., Coyle C., Creamer R.E., Fealy R. *et al.*, 2015. Making the Most of Our Land: Managing soil functions from local to continental scale. *Frontiers in Environmental Science*, 3, https://doi.org/10.3389/fenvs.2015.00081.

Serrano J., Vianey G., 2014. Patrimonialiser des activités agricoles pour banaliser la consommation d'espaces agricoles périurbains : réflexions à partir du cas de l'agglomération de Tours. *Géographie, économie, société*, 16 (3), 297-314, https://doi.org/10.3166/ges.16.297-314.

Shepherd G., 2000. *Visual Soil Assessment: Field guide for cropping*, Landcare Research, https://orgprints.org/30582.

Soenen B., Henaff M., Lagrange H., Lanckriet E., Schneider A., Duval R., Streibig J.-L., 2021. *Méthode Label Bas-Carbone Grandes Cultures* (version 1.0), 133 p., https://label-bas-carbone.ecologie.gouv.fr/la-methode-grandes-cultures.

Spake R., Barajas-Barbosa M.P., Blowes S.A., Bowler D.E., Callaghan C.T., Garbowski M. *et al.*, 2022. Detecting Thresholds of Ecological Change in the Anthropocene. *Annual Review of Environment and Resources*, 47, 797-821, https://doi.org/10.1146/annurev-environ-112420-015910.

Star S.L., Griesemer J.R., 1989. Institutional ecology, 'translations' and boundary objects: Amateurs and professionals in Berkeley's Museum of Vertebrate Zoology, 1907-39. *Social Studies of Science*, 19 (3), 387-420, https://doi.org/10.1177/030631289019003001.

Stevens A.W., 2018. The economics of soil health. *Food Policy*, 80, 1-9, https://doi.org/10.1016/j.foodpol.2018.08.005.

Styc Q., Lagacherie P., 2019. What is the best inference trajectory for mapping soil functions: An example of mapping soil available water capacity over Languedoc-Roussillon (France). *Soil Systems*, 3 (2), 17, https://doi.org/10.3390/soilsystems3020034.

Tafani C., Jouve J., 2023. *Penser l'usage agri-environnemental des sols : un enjeu pour un aménagement durable du territoire : L'exemple de la Balagne en Corse*, 17 p., https://doi.org/10.58110/ESTATE-TG55.

Te Wierik S.A., Gupta J., Cammeraat E.L.H., Artzy-Randrup Y.A., 2020. The need for green and atmospheric water governance. *Wiley Interdisciplinary Reviews-Water*, 7 (2), 20, https://doi.org/10.1002/wat2.1406.

Terrat S., Horrigue W., Dequietd S., Saby N.P.A., Lelièvre M., Nowak V. *et al.*, 2017. Mapping and predictive variations of soil bacterial richness across France. *Plos One*, 12 (10), 19, https://doi.org/10.1371/journal.pone.0186766.

Tomis V., Duparque A., Boizard H., 2019. Development of the "mini 3D soil profile" — A visual method derived from the "profil cultural". *Soil & Tillage Research*, 194, 8, https://doi.org/10.1016/j.still.2019.06.002.

Trevors J.T., 1984. Dehydrogenase activity in soil: A comparison between the INT and TTC Assay. *Soil Biology & Biochemistry*, 16 (6), 673-674, https://doi.org/10.1016/S0038-0717(96)00290-8.

Trompette P., Vinck D., 2009. Retour sur la notion d'objet-frontière. *Revue d'anthropologie des connaissances*, 3 (1), 5-27, https://doi.org/10.3917/rac.006.0005.

Tsiboe F., Tack J., 2022. Utilizing topographic and soil features to improve rating for farm-level insurance products. *American Journal of Agricultural Economics*, 104 (1), 52-69, https://doi.org/10.1111/ajae.12218.

Turnhout E., Hisschemöller M., Eijsackers H., 2007. Ecological indicators: Between the two fires of science and policy. *Ecological Indicators*, 7 (2), 215-228, https://doi.org/10.1016/j.ecolind.2005.12.003.

UE (Union européenne), 2023. *Proposal for a Directive on Soil Monitoring and Resilience*. COM_2023_416_final.

UE (Union européenne), 2024. *Proposal for a Directive on Soil Monitoring and Resilience*. 10236/24.

United Nations Convention to Combat Desertification, 2022. *The Global Land Outlook — Second edition — Land Restoration for Recovery and Resilience*, Bonn, UNCCD, https://www.unccd.int/sites/default/files/2022-04/UNCCD_GLO2_low-res_2.pdf.

Van Leeuwen C.C.E., Cammeraat E.L.H., de Vente J., Boix-Fayos C., 2019. The evolution of soil conservation policies targeting land abandonment and soil erosion in Spain: A review. *Land Use Policy*, 83, 174-186, https://doi.org/10.1016/j.landusepol.2019.01.018.

Veum K.S., Sudduth K.A., Kremer R.J., Kitchen N.R., 2017. Sensor data fusion for soil health assessment. *Geoderma*, 305, 53-61, https://doi.org/10.1016/j.geoderma.2017.05.031.

Villanneau E., Saby N., Jolivet C., Marot F., Maton D., 2008. Détection de valeurs anomaliques d'éléments traces métalliques dans les sols à l'aide du Réseau de Mesure de la Qualité des Sols. *Étude Gest. des Sols*, 15, 183 202.

Vincent Q., Auclerc A., Beguiristain T., Leyval C., 2018a. Assessment of derelict soil quality: Abiotic, biotic and functional approaches. *Science of the Total Environment*, 613, 990-1002, https://doi.org/10.1016/j.scitotenv.2017.09.118.

Vincent Q., Blanchart A., 2023. *Faciliter le transfert des résultats de la recherche en outils opérationnels — Thématique de la multifonctionnalité des sols*, Ademe, 97 p., https://librairie.ademe.fr/recherche-et-innovation/6788-le-bilan-de-la-recherche-sur-la-multifonctionnalite-des-sols.html.

Vincent Q., Leyval C., Beguiristain T., Auclerc A., 2018b. Functional structure and composition of Collembola and soil macrofauna communities depend on abiotic parameters in derelict soils. *Applied Soil Ecology*, 130, 259-270, https://doi.org/10.1016/j.apsoil.2018.07.002.

Voltz M., Arrouays D., Bispo A., Lagacherie P., Laroche B., Lemercier B. *et al.*, 2020. Possible futures of soil-mapping in France. *Geoderma Regional*, 23, 11, https://doi.org/10.1016/j.geodrs.2020.e00334.

Voltz M., Arrouays D., Bispo A., Lagacherie P., Laroche B., Lemercier B., Richer-de-Forges A.C., Sauter J., Schnebelen N., 2018. *La cartographie des sols en France: Etat des lieux et perspectives*, INRA, Paris, France, 112 p.

Wadoux A., Marchant B.P., Lark R.M., 2019. Efficient sampling for geostatistical surveys. *European Journal of Soil Science*, 70 (5), 975-989, https://doi.org/10.1111/ejss.12797.

Wahlhütter S., Vogl C.R., Eberhart H., 2016. Soil as a key criteria in the construction of farmers' identities: The example of farming in the Austrian province of Burgenland. *Geoderma*, 269, 39-53, https://doi.org/10.1016/j.geoderma.2015.12.028.

Wang M.E., Faber J.H., Chen W.P., Li X.M., Markert B., 2015. Effects of land use intensity on the natural attenuation capacity of urban soils in Beijing, China. *Ecotoxicology and Environmental Safety*, 117, 89-95, https://doi.org/10.1016/j.ecoenv.2015.03.018.

Weil R.R., Islam K.R., Stine M.A., Gruver J.B., Samson-Liebig S.E., 2009. Estimating active carbon for soil quality assessment: A simplified method for laboratory and field use. *American Journal of Alternative Agriculture*, 18 (1), 3-17, https://doi.org/10.1079/AJAA200228.

Weissgerber M., Roturier S., Julliard R., Guillet F., 2019. Biodiversity offsetting: Certainty of the net loss but uncertainty of the net gain. *Biological Conservation*, 237, 200-208, https://doi.org/10.1016/j.biocon.2019.06.036.

Wiesmeier M., Urbanski L., Hobley E., Lang B., von Lützow M., Marin-Spiotta E. *et al.*, 2019. Soil organic carbon storage as a key function of soils — A review of drivers and indicators at various scales. *Geoderma*, 333, 149-162, https://doi.org/10.1016/j.geoderma.2018.07.026.

Wolejko E., Jablonska-Trypuc A., Wydro U., Butarewicz A., Lozowicka B., 2020. Soil biological activity as an indicator of soil pollution with pesticides — A review. *Applied Soil Ecology*, 147, 13, https://doi.org/10.1016/j.apsoil.2019.09.006.

Zanella A., Ponge J.F., Jabiol B., Sartori G., Kolb E., Gobat J.M. *et al.*, 2018. Humusica 1, article 4: Terrestrial humus systems and forms — Specific terms and diagnostic horizons. *Applied Soil Ecology*, 122, 56-74, https://doi.org/10.1016/j.apsoil.2017.07.005.

Acronyms and abbreviations

ADEME Agence de la transition écologique – French Agency for Ecological Transition

AECM Agri-Environmental and Climate Measures

AFES Association française pour l'étude des sols – French National Soil Science Society

AFNOR Association française de normalisation – French Standardisation Association

AOC *Appellation d'Origine Contrôlée* – Controlled Designation of Origin

ARC Avoidance-Reduction-Compensation (sequence)

AWC Availavle Water Capacity

BDAT *Base de données des analyses de terre* – French Soil Test Database

BDETM *Base de données des éléments-traces métalliques* – French Trace Metal Database

CASH Comprehensive Assessment of Soil Health

CAP Common Agricultural Policy

CDTA *Cartes Départementales des Terres Agricoles* – Departmental Agricultural Land Maps

CEC Cation Exchange Capacity

CEMAGREF Centre d'étude du machinisme agricole et du génie rural des eaux et forêts – Centre for the Study of Agricultural Machinery and Rural Water and Forestry Engineering

Cerema Centre d'études et d'expertise sur les risques, l'environnement, la mobilité et l'aménagement – Centre for Studies and Expertise on Risks, the Environment, Mobility and Development

CGAAER Conseil général de l'alimentation, de l'agriculture et des espaces ruraux – General Council for Food, Agriculture and Rural Areas

CGEDD Conseil général de l'environnement et du développement durable (currently IGEDD) – General Council for the Environment and Sustainable Development

CNRS Centre national de la recherche scientifique – National Centre for Scientific Research

CPF Connaissance pédologique de la France – French Program for Soil Mapping at Medium Scales (1/100 000 and 1/50 000)

CSLF Credibility-Salience-Legitimacy-Feasibility

CSM Conventional Soil Mapping

DDT Direction départementale des territoires – Departmental Directorate for Territories

DEPE Direction de l'expertise scientifique collective, de la prospective et des études – Directorate for Collective Scientific Assessment, Foresight and Advanced Studies (INRAE)

DNA Deoxyribonucleic Acid

DPSIR Drivers-Pressures-State-Impacts-Responses

DRAAF Direction régionale de l'agriculture, de l'alimentation et de la forêt – Regional Directorate for Agriculture, Food and Forestry

DSA Digital Soil Assessment

DSM Digital Soil Mapping

Ece Electrical conductivity

EEA: European Environment Agency

ENAF *Espaces naturels, agricoles et forestiers* – Natural, Agricultural and Forest Areas

ENS *Espace naturel sensible* – Sensitive Natural Areas

EPCI Etablissement publics de coopération intercommunale – Public Establishment for Inter-municipal Cooperation

EU European Union

EUSO European Union Soil Observatory

FAO Food and Agriculture Organisation

GAEC Good Agricultural and Environmental Conditions

GIS Geographic Information System

GIS Sol Groupement d'intérêt scientifique sur les sols – Scientific Interest Group on Soil

IGCS Inventaire, gestion et conservation des sols – Soil Inventory, Management and Conservation

IGN Institut national de l'information géographique et forestière – National Institute of Geographic and Forest Information

INRAE Institut national de recherche pour l'agriculture, l'alimentation et l'environnement – National Institute for Agricultural Research, Food and the Environment (since 2020)

SQI Soil Quality Index

ISO International Standardisation Organisation

JRC Joint Research Centre (European Union)

LBC Label Bas Carbone – Low Carbon Label

LULUCF Land Use, Land Use Change and Forestry

MASAF Ministère de l'agriculture, de la souveraineté alimentaire et de la forêt – Ministry of Agriculture, Food Sovereignty and Forestry

MAOC Mineral-Associated Organic Carbon

MDS Minimum Data Set

MNHN Museum national d'histoire naturelle – National Museum of Natural History

MTEECPR Ministère de la transition écologique, de l'éneregie, du climat et de la prévention des risques – Ministry of Ecological Transition, Energy, Climate and Risk Prevention

NASA National Aeronautics and Space Administration (United States)

OCS GE Observatoire de la couverture des sols à grande échelle – Large-scale Land Cover Observatory

OFB Office français de la biodiversité – French Office for Biodiversity

ONB Observatoire national de la biodiversité – French National Biodiversity Observatory

ONF Office national des forêts – French National Forestry Office

NSP National Strategic Plan

OM Organic Matter

OPVT Observatoire Participatif des Vers de Terre – Participatory Earthworm Observatory

PADD *Projet d'aménagement et de développement durables* – Sustainable Development and Planning Project

PAH Polycyclic Aromatic Hydrocarbons

PBC Potential Buildability Coefficient

PCA Principal Component Analysis

PCB Polychlorinated Biphenyls

PEAN *Périmètres de protection des espaces naturels et agricoles péri-urbains* – Protection Perimeters for Natural and Agricultural Areas in Peri-urban Areas

PFAS Per- and Polyfluoroalkyl Substances

PLFA Phospho Lipid Fatty Acid Analysis

PLU *Plan local d'urbanisme* – Local Urban Development Plan

PLUi *Plan local d'urbanisme intercommunal* – Inter-municipal Local Urban Development Plan

POC Particulate Organic Carbon

PPRN *Plan de Prévention des Risques Naturels* – Natural Risk Prevention Plan

PRA *Petites Région Agricole* – Small Agricultural Region

PSR Participatory Science and Research

PTF Pedotransfer Function

RMQS Réseau de mesure de la qualité des sols – French Soil Quality Monitoring Network

RRP *Référentiel regional pédologique* – Regional Soil Reference System

RUSLE Revised Universal Soil Loss Equation

Scot *Schéma de cohérence territoriale* – Territorial Coherence Scheme

SER Society for Ecological Restoration

SDGs Sustainable Development Goals

SMAF Soil Management Assessment Framework

SMRL Soil Monitoring and Resilience Law

SNB *Stratégie nationale pour la biodiversité* – National Biodiversity Strategy

SOC Soil Organic Carbon

SOM Soil Organic Matter

SQI Soil Quality Index

SMU Soil Mapping Unit

STU Soil Typological Unit

TRL Technology Readiness Level

UAA Utilised Agricultural Area

UN United Nations

USA United States of America

USDA United States Department of Agriculture

TMM Trace Metal and Metalloid
VESS Visual Evaluation of Soil Structure
VSA Visual Soil Assessment
WHC Water Holding Capacity
WoS Web of Science
WRB World Reference Base for Soil Resources
ZAP *Zone agricole protégée* – Protected Agricultural Area

Working group

Scientific leads

Isabelle COUSIN (RD): INRAE – AgroEcoSystem*, UR Info&Sol (State, functioning, soil monitoring, evaluation of services and soils, impacts, soil health and quality, environmental data management), Orléans. *Soil physics, Ecosystem functions and services.*

Maylis DESROUSSEAUX (AP): Paris School of Urban Planning (Urban Planning and Development), Champs-sur-Marne. *Environmental law, Soil protection.*

Scientific experts

Denis ANGERS (RD): Agriculture and Agri-Food Canada (AAFC), Quebec Research and Development Centre, Quebec City, Canada. *Soil science, Agronomy, Soil carbon.*

Laurent AUGUSTO (RD): INRAE – ECODIV*, UMR ISPA (Soil-Plant-Atmosphere Interactions) (effects of global change on ecosystems, material and energy transfers), Villenave, France. *Forest soil ecology, Soil functions.*

Jean-Sauveur AY (R): INRAE – EcoSocio*, UMR CESAER (Centre for Applied Rural Economics and Sociology in Agriculture and Rural Areas), Dijon, France. *Land economics, Land prices and use, Public policy.*

Adrien BAYSSE-LAINÉ (R): CNRS – UMR Pacte (Geography, Planning, Urbanism, Political Science, Sociology), Grenoble, France. *Geography of law and social geography of the environment, Agricultural and forest land, Property-management-use relationships.*

Philippe BRANCHU (R): Cerema, TEAM (Urban Hydrology, Environmental Sciences), Trappes, France. *Geochemistry, Nature-based Solutions (water, soil), Urban Soils.*

Alain BRAUMAN (RD): IRD, UR Eco&Sol (Functional soil ecology, role of organisms in biogeochemical cycles), Montpellier, France. *Soil ecology.*

Nicolas CHEMIDLIN PRÉVOST BOURÉ (Pr): Institut Agro Dijon, UMR Agroecology (Soil Microbiology, Agronomy, Plant-Microorganism Interactions, Genetics), Dijon, France. *Soil Microbial Ecology, Biogeography, Impact of Agricultural Practices.*

Claude COMPAGNONE (Pr): Institut Agro Dijon, UMR CESAER (Centre for Applied Rural Economics and Sociology in Agriculture and Rural Areas), Dijon, France. *Sociology.*

Raphaël GROS (Pr): Aix-Marseille University, UMR IMBE (Mediterranean Institute of Marine and Terrestrial Biodiversity and Ecology) (Biodiversity and socio-ecological systems), Marseille, France. *Soil ecology and restoration.*

* INRAE departments: AgroEcoSystem: Agronomy and environmental sciences for agroecosystems; ECODIV: Ecology and biodiversity; EcoSocio: Economics and social sciences. RD: Research Director; AP: Associate Profesor; R: Researcher; RE: Research Engineer; Pr: Professor.

Carole HERMON (Pr): Toulouse Capitole University, UR IEJUC (Institute for Legal Studies in Urban Planning, Construction and the Environment) (Environmental law, urban planning law, real estate law, rural law, property law), Toulouse, France. *Environmental law, Agriculture, Litigation.*

Catherine KELLER (Pr): Aix-Marseille University, UR Cerege (Centre for Research and Teaching in Environmental Geosciences) (Geosciences, sustainable development, risks, climate, current environments, formation and deformation of the Earth's surface), Aix-en-Provence, France. *Biogeochemistry, Contamination-(phyto)remediation, Urban soils.*

Germain MEULEMANS (R): CNRS – UMR Centre Alexandre-Koyré (Social and Cultural History of Science, Knowledge and Technology), Paris, France. *Environmental anthropology, Science and Technology Studies (STS).*

David MONTAGNE (AP): AgroParisTech – UMR EcoSys (Functional Ecology and Ecotoxicology of Agroecosystems), Saclay, France. *Pedology, Pedogenesis, Ecosystem services.*

Guénola PÉRÈS (AP): Institut Agro – UMR SAS (Soil Agro Hydrosystem Spatialisation) (Agriculture-Environment Interactions), Rennes, France. *Functional Soil Ecology, Bioindicators, Earthworms, Ecosystem Functions and services.*

Nicolas SABY (RE): INRAE – AgroEcoSystem*, UR Info&Sol (Soil status, functioning, monitoring, evaluation of services and soils, impacts, soil health and quality, environmental data management), Orléans, France. *Geostatistics, Monitoring, Modelling, Spatial analysis, Statistical information system.*

Jean VILLERD (RE): INRAE – AgroEcoSystem*, UMR LAE (Agronomy and Environment Laboratory) (Functioning of plants, crops, agricultural production systems and territories), Nancy/Dijon, France. *Modelling, Data analysis, Software development.*

Cyrille VIOLLE (RD): CNRS – UMR CEFE (Centre for Functional and Evolutionary Ecology) (Biodiversity, ecology, evolutionary biology), Montpellier, France. *Functional ecology, Biodiversity, Eco-evolutionary dynamics, Biogeography.*

Scientific experts – occasional contributors

Bertrand LAROCHE (RE): INRAE – AgroEcoSystem*, UR Info&Sol (Soil status, functioning, monitoring, assessment of services and soils, impacts, soil health and quality, environmental data management), Orléans, France. *Pedology, Coordination of the national Soil Inventory, Management and Conservation (IGCS) programme.*

Emmanuelle VAUDOUR (RD): INRAE – UMR EcoSys (Functional Ecology and Ecotoxicology of Agroecosystems), Saclay, France. *Pedology, Remote Sensing, Spatial analysis, Cartography.*

Project managers

Julie ITEY (IR): INRAE – DEPE, Paris, France. *Contextual elements, case law study.*

Marie-Caroline BRICHLER (IR): INRAE – Info&Sol, Orléans, France. *Trial phase.*

Claire FROGER (IR): INRAE – Info&Sol, Orléans, France. *Trial phase.*

* INRAE departments: AgroEcoSystem: Agronomy and environmental sciences for agroecosystems; ECODIV: Ecology and biodiversity; EcoSocio: Economics and social sciences. RD: Research Director; AP: Associate Profesor; R: Researcher; RE: Research Engineer; Pr: Professor.

Documentation

Sybille DE MARESCHAL: INRAE – DipSO (Open Science Directorate, Astra team), Bordeaux, France.
Virginie LELIÈVRE: INRAE - AgroEcoSystem*, Avignon, France.

Graphic design

Sacha DESBOURDES: INRAE – AgroEcoSystem*, Orléans, France. *Graphic designer*.
Maëlle JOLY: freelance graphic designer. *Illustration*.

Directorate for Collective Scientific Assessment, Foresight and Advanced Studies (DEPE)

Sophie LEENHARDT. Project lead, writing.
Claire MEUNIER. Project management support.
Kim GIRARD. Communication and administrative management support.
Sandrine GOBET. Logistics and administrative management support.
Isabelle SAVINI. Editorial support.

Monitoring committee

Members: Ademe (Miriam Buitrago, Antoine Pierart), OFB (Nolwenn Bougon, Kathleen Monod), MTEECPR/CGDD (Véronique Antoni then Benjamin Trochon, Irénée Joassard), MASAF/DGER (Patricia Laville), INRAE Environment Deputy (Pierre Renault) and INRAE DEPE (Guy Richard).

Stakeholder Advisory Committee

Participating organisations: Association for Technical Coordination in Agriculture (ACTA – Association de coordination technique agricole), French Agronomy Association (AFA – Association française d'agronomie), Agricultural, Forestry and Environmental Land Management (AFAFE – Aménagement foncier Agricole, forestier et environnemental), French National Soil Science Society (Afes – Association française pour l'étude des sols), Association for the Promotion of Sustainable Agriculture (APAD – Association pour la promotion de l'agriculture durable), Permanent Assembly of Chambers of Agriculture (APCA, now Chambres d'Agriculture France), Association of French Regions (ARF – Association des regions de France),

* INRAE departments: AgroEcoSystem: Agronomy and environmental sciences for agroecosystems; ECODIV: Ecology and biodiversity; EcoSocio: Economics and social sciences. RD: Research Director; AP: Associate Profesor; R: Researcher; RE: Research Engineer; Pr: Professor.

Caisse des Dépôts et Consignations (CDC) Biodiversité, National Centre for Forest Ownership (CNPF – Centre national de la propriété forestière), National Federation of Organic Agriculture in the Regions of France (FNAB – Fédération nationale d'agriculture biologique des régions de France), National Federation of Scot (FN-SCoT), France Nature Environment (FNE), Foundation for Nature and Mankind (FNH – Fondation pour la nature et l'Homme), National Federation of Land Development and Rural Settlement Companies (FN-SAFER – Sociétés d'aménagement foncier et d'établissement rural), Humanité et biodiversité, Laboratory for Innovative Land and Territorial Initiatives (LIFTI – Laboratoire d'initiatives foncières et territoriales innovantes), League for the Protection of Birds (LPO – Ligue pour la protection des oiseaux), Terre de Liens-Fédération, Union of Environmental Consultants and Engineers (UCIE – Union des consultants et ingénieurs en environnement), Professional Union of Ecological Engineering (UPGE – Union professionnelle du genie écologique).

The authors would like to thank the contributors and reviewers of the report and this condensed report.

Cover photograph: photo © INRAE – Bertrand NICOLAS – 2021
Graphic designer: © S. Desbourdes – INRAE
Editing: Mickaël Legrand
Layout and computer graphics: Hélène Bonnet, Studio 9
Printed for you by Libri Plureos GmbH (Germany)
Legal deposit: August 2026

www.ingramcontent.com/pod-product-compliance
Lightning Source LLC
LaVergne TN
LVHW010723061126
851857LV00014B/1319